11th International Robotic Sailing Conference 2018

CEUR Workshop Proceedings Volume 2331

Southhampton, United Kingdom
31 August – 1 September 2018

Editors:

Sophia M. Schillai
Nicholas Townsend

ISBN: 978-1-7138-2912-6

Printed from e-media with permission by:

Curran Associates, Inc.
57 Morehouse Lane
Red Hook, NY 12571

Some format issues inherent in the e-media version may also appear in this print version.

Printed with permission by Curran Associates, Inc. (2021)

For permission requests, please contact the Editors at the address below.

University of Southampton - Faculty of Engineering
Boldrewood Campus
Burgess Road
Southampton, SO16 7QF
United Kingdom

http://ceur-ws.org/Vol-2331/

Additional copies of this publication are available from:

Curran Associates, Inc.
57 Morehouse Lane
Red Hook, NY 12571 USA
Phone: 845-758-0400
Fax: 845-758-2633
Email: curran@proceedings.com
Web: www.proceedings.com

Sophia M. Schillai
Nicholas Townsend (Eds.)

ROBOTIC SAILING 2018

Proceedings of the 11th International Robotic Sailing Conference

Southampton, United Kingdom, August 31, 2018

ROBOTIC SAILING 2018

Proceedings of the 11th International Robotic Sailing Conference

Preface

In the words of Kenneth Grahame "*there is nothing–absolutely nothing–half so much worth doing as simply messing about in boats.*" With that in mind, the 2018 International Robotic Sailing Conference (IRSC) and World Robotic Sailing Championship (WRSC), hosted by the University of Southampton, was no exception!

This year's event, not only marked the 10th anniversary of WRSC & IRSC, but also coincided with a long awaited success: the first robotic sailing vessel to successfully complete the Microtransat challenge, the first crossing of the Atlantic Ocean since the competition conception in 2005. The WRSC event, which ran from 26th August through to the 1st September 2018 was held at the historical Calshot Hanger – a split of land extending into the Solent, just outside Southampton. An area renowned for sailing, the hovercraft, the Spitfire and the Titanic. This provided a spectacular backdrop for the 5 day competition, with a variety of Ocean liners, cruise ships, container ships, ferries and yachts passing by as they came into and out of Southampton!

With industry and research in the maritime sector increasingly focusing on algorithms for planning and control of vehicles as part of collaborative tasks, WRSC and the development of autonomous boats through friendly competition continue to become more prominent and important. Inspired by the collaborative missions of surface and underwater vehicles demonstrated by the National Oceanography Centre and ASV global, the difficulty of the area scanning challenge was increased by making it a collaborative task. This lead to, in one afternoon, an impressive 28 000+ square metres being scanned at a resolution of 4 m x 4 m.

Following the World Robotic Sailing Championship, the International Robotic Sailing Conference provided a forum to further discuss and share both WRSC vessel designs (including sensor developments, modelling and control methods) as well as discuss the development of maritime autonomous vehicles more widely. The 2 day conference, hosted at the University of Southampton Boldrewood Campus, included presentations of the eleven peer reviewed papers (published here). In addition a series of lighting talks, discussion sessions, and further discussions during the conference dinner and tour of the National Oceanography Centre (NOC) inspired the competitors and researchers for next year's event.

The editors would like to thank all authors, the Program Committee, the Southampton volunteers, the University of Southampton, the National Oceanography Centre, ASV global and all other sponsors whose contribution made the 2018 IRSC and WRSC possible. Thank you!

Organizing Committee

Sophia M. Schillai, University of Southampton, United Kingdom
Nicholas Townsend, University of Southampton, United Kingdom

Program Committee

Nicholas Townsend, University of Southampton, United Kingdom
Sophia M. Schillai, University of Southampton, United Kingdom
Sébastien Lemaire, University of Southampton, United Kingdom
Yu Cao, University of Southampton, United Kingdom
Fabrice Le Bars, ENSTA Bretagne, France
Cedric Pradalier, GeorgiaTech Lorraine, France
Erik Maehle, University of Luebeck, Germany
Thomas Kluyver, University of Southampton, United Kingdom
Nuno Cruz, Universidade do Porto, Portugal
Antonio C. Dominguez-Brito, Universidad de Las Palmas de Gran Canaria, Spain
Anna Friebe, Åland University of Applied Sciences, Finland
Paul Miller, US Coast Guard Academy, USA
Colin Sauze, Aberystwyth University, Wales, United Kingdom
Benedita Malheiro, Instituto Superior de Engenharia do Porto, Portugal
Oren Gal, Technion Israel Institute of Technology, Israel
Ole Blaurock, Fachhochschule Lübeck, Germany
Kostia Roncin, ENSTA Bretagne, France
Jorge Cabrera, Universidad de Las Palmas de Gran Canaria, Spain
Paul Miller, USCGA, USA
Michael Schukat, NUI Galway, Ireland
Diedrich Wolter, University of Bamberg, Germany
Noureddine Bouhmala, Borre, Norway
Alexander Schlaefer, Hamburg University of Technology, Germany
Alex Laun, United States Naval Academy, USA
Artur Lidtke, University of Southampton, United Kingdom
Bradley Bishop, United States Naval Academy, USA
Manuel Silva, ISEP-IPP and INESC TEC CRIIS, Portugal

Contents

Part I

Vessel Development & Modelling

Development and Test of an Open Source Autonomous Sailing Robot with Accessibility, Generality and Extendability

Ulysse Vautier
ulysse.vautier@plymouth.ac.uk

Jian Wan
jian.wan@plymouth.ac.uk

Ming Dai
y.dai@plymouth.ac.uk

Christophe Viel
christophe.viel@plymouth.ac.uk

Robert Hone
robert.hone@plymouth.ac.uk

School of Engineering, University of Plymouth, Plymouth

Abstract

This paper introduces the design of an open source autonomous sailing robot with emphasis on its accessibility, generality and extendability. To meet such requirements, a generic control box connecting an Arduino board and a Raspberry Pi computer was tailor-made so as to host the robot operating system (ROS) and to interact with versatile sensors and actuators. The goal of such a project is to create an accessible, generic and expendable platform of autonomous sailboats for wider education and research publicity and engagement. Autonomous sailing test for the developed sailboat was also conducted to validate its design.

1 Introduction

Over the last two decades, there are growing interests in developing autonomous sailing robots. Different groups have already made fully autonomous sailboats, ready to deploy for data gathering purposes and almost able to cross the transatlantic ocean in the Microtransat challenge (Meinig et al., 2015; Ghani and Hole, 2014). Sailboat research projects are also growing, most of which are particularly structured for specific problems (Naveau et al., 2013; Rathour et al., 2017; Silva Junior et al., 2016; O'Hara, 2017) or others for generic research (Domínguez-Brito et al., 2015; Miller et al., 2014; Wirz et al., 2015; Plumet et al., 2015). There are also various kinds of software, hardware or mechanical design projects (Santana-Jorge et al., 2017; Neal, 2006; Lam et al., 2016). While the motives of building such robots are multiple with no wide-spread applications for autonomous sailing technologies, the research in this field is still in its early stage compared to other autonomous vehicles such as cars or drones.

An autonomous sailboat is interesting in that it can use all available energy around it to complete its mission i.e. wind, solar and wave energy (Liu et al., 2016). This makes it an important subject for long-duration applications such as oceanography, surveillance and reconnaissance. Moreover, such robots can play a significant role in solving nowadays environmental and ecological problems. The energy needed to operate a sailboat is relatively small in comparison to other autonomous vehicles (Cruz and Alves, 2008). Nevertheless, the market

In: S. M. Schillai, N. Townsend (eds.): Proceedings of the International Robotic Sailing Conference 2018, Southampton, United Kingdom, 31-08-2018

and the commercial applications are yet to be explored to bring greater interest to it. In that context, there is a need to develop a modular sailing robot platform with accessibility, generality and expandability so as to facilitate the development of such robots with more focus on their control and applications.

Most projects of autonomous sailboats are open source. While most of them explain the hardware specifications and some explain the software designs, it is still hard for a researcher to make their own sailboat in a short amount of time and to focus on control algorithms or applications. This asks for an accessible hardware, already available in most research labs and more importantly a friendly software architecture which makes it straightforward for any user to test control algorithms.

In the same way 3D printers are designed, fully open-sourced and with accessible hardware, a control box was made for highly generic hardware and software structures. That way, the control box can be put in any RC (Radio-Controlled) sailboat hull, connect to sensors and servo-motors and transform it into an autonomous sailboat. The software architecture has to be flexible so that in couples of lines of code, the sailboat can respond and act according to the instructions. So this paper will address the development of an accessible, generic and expendable autonomous sailboat with particular emphasis on the aspect of control hardware and software, which is in the format of a control box.

2 Accessible Hardware Configuration

The hardware architecture is similar to other sailboats. To stay low on budget and to use reliable resources, most projects use Arduino boards and Raspberry Pis, which can be easily accessed by most robotics laboratories (Krauss, 2016; Lazarin and Pantoja, 2015; Miller et al., 2014). The hardware structure is also mostly the same (Lam et al., 2016; Plumet et al., 2015), it consists of a low-level computer and a high-level computer. In this case, the Arduino acts as a low-level controller acquiring data from sensors and acting on the actuators and the Raspberry Pi acts as a high-level controller with more intensive algorithms as shown in Figure 1. Making a specific purpose board can lower the price and simplify the architecture (Alvira et al., 2013; Cabrera-Gámez et al., 2013), but it would decrease the accessibility of such a hardware. Others would use a bigger computer instead of the Raspberry Pi to have more resources for the algorithms (Naveau et al., 2013; Lam et al., 2016) but this will increase the cost and reduce accessibility. The newest Raspberry Pis are good enough to process complex control algorithms (Benchoff, 2016; Larabel, 2018). It would also be possible to use the Raspberry Pi alone. Adding low-level microcontrollers provide an extra security and free valuable resources of the high-level computer for main control tasks.

Along with the technology development, the price for Arduino boards and Raspberry Pis will continue to decrease while their performance will continue to improve, which makes such a hardware architecture more attractive in the long term.

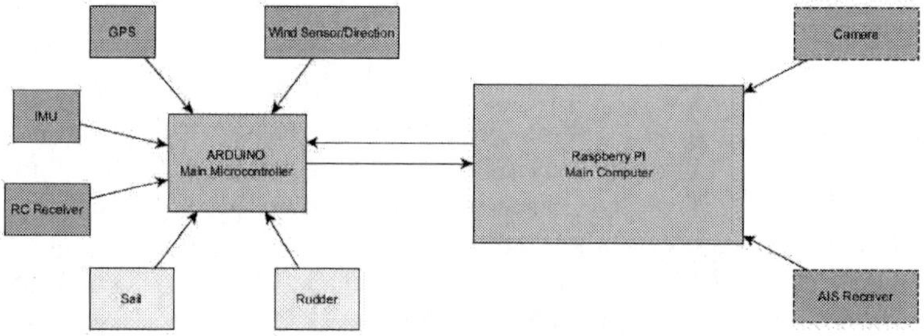

Figure 1: Hardware Architecture.

2.1 Specifications and Architecture

The control box was designed to comply with as many standards as possible. While the boards presented here comply with our configurations, changes can be made to accept other configuration of power necessities. The sailboats are powered with a 2S (Two cells) LiPo battery which outputs a maximum of 8.4V (4.2V per cells) and nominal at 7.4V (3.7V per cells). We set our requirements to at least two hours of continuous operation without any other energy input such as solar panels. From all the sensors and actuators the sailboats have, we are at about 2A usage at full use. This means we must have at least a 4000mAh battery. Of course, this also depends on other factors such as the boat size, weight and usage. For standardization, it is considered that 3S (maximum 12.6V) or even 4S (maximum 16.8V) batteries can be used.

Particular sensors or actuators are not discussed in this paper. But based on RC models of sailboats and expertise of users, a standard as generic as possible was designed. As such, a 3 pin connector was considered for small servomotors connection. Other connectors were chosen arbitrarily for robustness while keeping standard pin configurations. Those configurations are to be discussed further. For a sailboat to be autonomous, it needs at least a way to know its position, heading and wind direction. We will use a GPS, IMU and a wind direction sensor in our examples. This control box is designed to be able to connect with other sensors. Concerning actuators, it has to control the sail and the rudder.

2.2 Custom Interface Board

While the platforms chosen are accessible, they are development platforms and are unfit for industrial applications. Pin headers are the connectors in both boards. They are not suitable for robust and waterproof connections. To solve this problem, a custom board is made to have better industry-grade connectors. This custom board decreases the accessibility for researchers but it highly improves the robustness of the control box. Making such a board can hinder the accessibility with some laboratories which do not have the facilities to make such a PCB board. We suggest the use of third-parties industries, accessible to anybody, to make those boards.

This custom board is connected to the Arduino and the Raspberry Pi. It manages the power and extracts all connectors from the Arduino to sensors and actuators. It is also designed to contain all the boards in a small sized box and be mechanically robust. The following sections will describe in detail the choices in the different parts of the boards. All the details and documents for DIY can be found in the *Meca* Repository of (Sailboat, 2018).

2.2.1 Power Management

The power management is simple and efficient. The unit manages the energy supply for sensors, actuators and the Raspberry Pi. It is simply made of regulators for each group of loads. We use switching regulators because of its high efficiency. These regulators can input a large range of voltage (6.5-32V) and output a constant voltage (5V). They are used for the actuators and the Raspberry Pi. One linear regulator can manage the power supply to multiple sensors because of the low energy consumption of sensors. While these regulators were chosen based on our configuration, the footprints on the custom boards are standard, making it possible to solder any regulators.

Heat and energy is a problem for underwater and surface vehicle. Using linear regulators or regulators with low efficiency can hinder the boat capacity. Switching regulators are used with efficiency up to 96% to lower the heat generated. The Raspberry Pi is the biggest consumer of energy. In a linear regulator, the ballast lowers the energy by outputting heat. This would end up wasting $(8.4V - 5V) \times 2A = 6.8W$ of energy while a switching regulator would waste only $5V \times 2A \times (\frac{1}{0.96} - 1) = 0.42W$ (considering 96% efficiency) for the Raspberry Pi.

2.2.2 Connectors and Communication

Connectors play a big part in robustness for robots. In mobile robots, vibration and movement can remove the cables from the pins. Robust, industry-grade connectors that are easy to plug in and out frequently was necessary for sensors. JST connectors were chosen purposely for the latter, based on usage experience and industry standards. Concerning the battery connections, two pin holes are placed on the custom board on which any wire connectors can be soldered. In our case, our battery had HXT 4mm connectors and as such HXT wire adapters were soldered, as shown in Figure 2. Finally, the custom boards also have a special compartment for radio receivers, for which a row of 3 pin holes is made available. Up to 6 channels are pluggable.

There are some standards in connection and communication protocols for most sensors and actuators. Most notable communication protocols range from Serial communication, I2C protocol to Analog signal and Digital interrupts. Each of these communications come with a standard wiring. Sensors have 2 wires for power - Ground and Power (5V) - and the rest for communication - RX, TX for Serial, SDA, SCL for I2C, or 1 pin for analog and digital signal. Of course, other protocols and wirings exist but those are the most common.

This custom board accepts all of those different connectors and communications and allows access to the corresponding Arduino pins with mechanical connectors. Once connected, the software part arranges the connection inside the system through a configuration file.

2.2.3 Waterproofing

One of the challenges in sailboats is waterproofing all the electronic components. Several 3D printed boxes were made for the sensors and the control box, all available online at the *Meca* repository of (Sailboat, 2018). A key problem arising from waterproofing is heat. Any electronic components generate heat. An enclosed system in plastic or wood insulates it and keep the heat inside, which may cause the components to overheat. For low-power sensors such as GPS or IMUs, plastic is enough to dissipate the heat. But for heat-generating components such as the main computer, a high dissipating material is needed. For this case, a side of the box (close to the main computer) is metal.

The 3D printed case for the boards' stack is a frame-like box which leaves open the connectors. While the box is not fully waterproof, it is enough to protect the boards from the small amount of water coming inside the boat. Batteries were chosen to be ROAR approved, meaning they are approved for racing. These batteries are hard-cased and protected against high impact (Rules, 2013). The hard case in itself provides a weatherproof package for batteries. The hulls of the boat are sealed as much as possible to waterproof the whole sailboat.

2.3 Hardware

Figure 2 shows the board and its components. Two boards were made to form the overall structure. The overall board stack dimensions are 70x130x40mm. It was designed to be placed in our 1m-long boat[1]. Table 1 shows the overall components used and their specifications. A complete detailed BOM (Bill of Material) can be found in the *Meca* repository of (Sailboat, 2018). It is based on the high prices of each component, for instance, the Arduino Mega can be found at a lot cheaper cost. The cost of the custom boards is at £40.1 using a third-party service to print the circuit boards. The biggest cost comes from the switching regulators. The remaining cost of the whole sailboat would be on the hull, the sensors, actuators and battery.

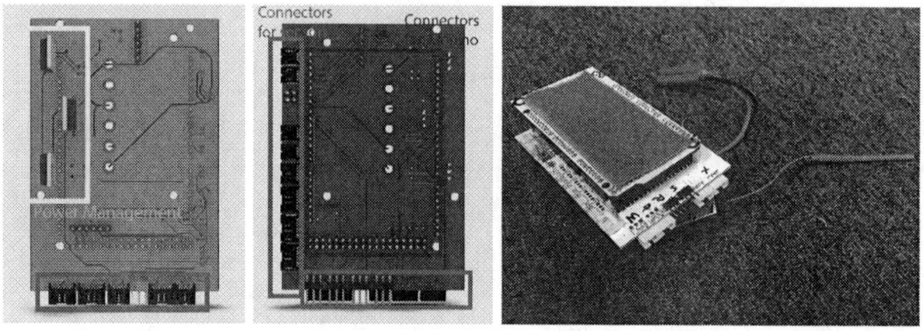

Figure 2: Back and front of the custom board and the assembled control box.

Table 1: Hardware Specifications and Approximate Price of the Control Box (without sensors and motors)

Part	Description	Total Price (£)
Board Components	Resistors, Capacitors, Inductors, diodes and regulators	19.1
Connectors	Industrial Grade Connectors - Arduino and Raspberry Pi Connectors	12
Circuit Board	Third-Party Service from SeeedStudio	8
Screws and Spacers	Mechanical constraints - M3	1
Arduino Mega	Low-level Microcontroller Board	28
Raspberry Pi 3 Model B	High-level SBC Computer	32
Total		100.1

3 Generic Software Architecture

Sailboats come in different formats, small or long, one or two rudders, winch motors for sails or direct actuation. As the hardware can change among different sailboats, the software architecture should be able to adapt to

[1]Length Over All

those varying configurations. In a similar way with 3D printers that built upon the marlin firmware (Marlin,), a configuration file is used to configure the customized hardware specifications. Not only does the hardware configuration change but different brands and different protocols of sensors and actuators are used. This is solved by using a component-based architecture, highly used in the video game industry and robotics (Gregory, 2014; Elkady and Sobh, 2012; Santana-Jorge et al., 2017). It simplifies the development of libraries of different components while maintaining uniformed standard and compatibility with the overall system.

Concerning the Raspberry Pi, no configuration is needed. The focus is on the usability and interfaces for users to add their own controllers. **ROS** (Robot Operating System) has been increasingly used for robotics research and development (Cousins, 2011). Using ROS as the base of the software architecture, the project can benefit from existing work done by other researchers in robotics and be able to share its specific outcomes with the community as well. With **ROS**, a software architecture has been done for users to easily and rapidly integrate their controllers into the robot.

3.1 Arduino Programming

The Arduino is the low-level controller. It acquires all the data from different sensors and acts upon the actuators. Once all the data is updated, it sends them to the main computer and also receives the instructions from it. The code and the wiki for installation instructions can be found in *Arduino Interface* repository of (Sailboat, 2018).

A Component-Based architecture is exploited to be as generic as possible with regards to the hardware. Such architecture fully uses the **OOP** (Oriented-Object Programming) aspect of C++, the language used for Arduino. The principle is based on the semantics of the architecture. We have a main class, the *Sailboat*, which contains component objects : *Sensors*, *Actuators* and *Basic Controllers*. Those components have their own methods to conduct certain tasks. *Sensors* have the common methods of acquiring data from sensors and sending them to the main computer. *Actuators* have the common method of acting upon the actuators to the desired set-point. Finally, *Basic Controllers* have the common methods of controlling, taking the data from sensors and acting upon the actuators. The implementation of those methods might differ among different products but the inputs and outputs are the same. Using **OOP**, you can derive from these components class and would only need to implement the methods. The *Sailboat* class would automatically call the right methods depending on the derived component class. This architecture makes it simple for any user, using a new sensor or actuator, to implement the methods and stay compatible with the sailboat hardware configuration.

The *Basic Controllers* are low-level controllers such as rudder control, sail control or heading control. This enables the main computer to send different inputs to the Arduino, depending on the controller, e.g. A Potential Field controller might output a heading while a Line Follower controller might directly act upon the rudder without adjusting the sail.

On top of those basic controllers are security controllers such as the RC controller and the Return Home controller. They will be discussed further in section 3.3 on safety.

Figure 3 shows a UML (Unified Modeling Language) of the different classes in the Arduino architecture. For the optimal sail angle and the rudder angle, we are constructing our code on (Jaulin and Le Bars, 2012) for our basic controllers. You can see the implementation of all the methods at *Arduino Interface* repository of (Sailboat, 2018).

3.2 Raspberry Pi Programming and ROS

The Raspberry Pi is the main computer in this architecture. It is loaded with a lightweight *Ubuntu* Operating System. Its goal is to calculate the complex control algorithms using the data sent by the Arduino and act on the actuators by commanding the Arduino. In both the Arduino and the Raspberry Pi, the architecture is made so that any users will be able to focus on the algorithm and not the integration of it. The architecture also uses the **OOP**. **ROS** is compatible with Python and C++ languages and the same architecture is used for both languages. The code related to the Raspberry Pi can be found on *ROS* repository of (Sailboat, 2018).

This architecture is different from the Arduino software architecture in that it focuses on usability and ergonomics for users. The goal of such an architecture is to free the user from integration problems and focus only on the important part of the research, which is the control of the sailboat.

For that reason, the architecture automatically takes over communication tasks with the Arduino, acquiring and parsing the data from it and making it available to the user. All the user needs to do is to implement two methods : a setup method to initialize all the variables and a control method where the algorithm has to be implemented, which sends the results to the Arduino.

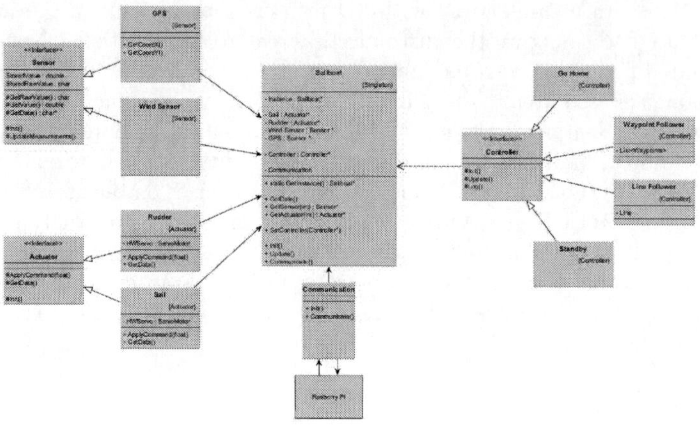

Figure 3: UML of the Component-Based Architecture.

The newest Raspberry Pis come with a WiFi chip. This enables WiFi communication with another computer without cables. When it boots up, it creates a hotspot, it is then possible to connect to it, SSH (Secure Shell) into the Raspberry Pi and controls it directly from a laptop.

ROS is used for multiple reasons. It is first highly used by the robotics community around the world and some users might consider it standard. It also helps in making robust protocol communication with the Arduino and among different components of the sailboat. Finally, it already has various libraries/applications (called *nodes* in **ROS**) used in robotics. For the latter to work, a certain standard had to be met while working on the architecture to make it compatible with most nodes, particularly on the message structures of each sensor.

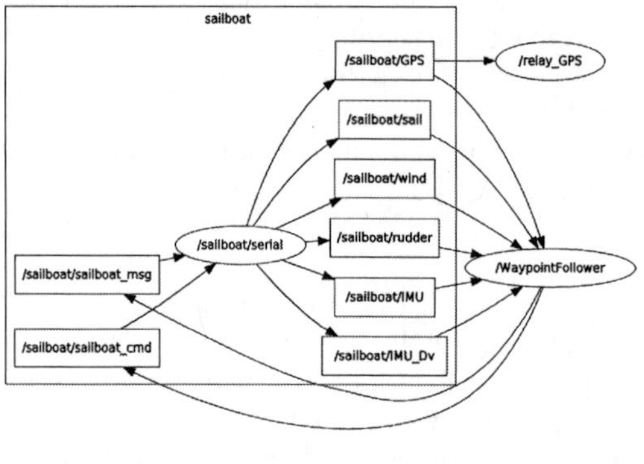

Figure 4: A graph of ROS topics and nodes when applying a waypoint-following control.

All the topics concerning the sensors are published by the Arduino. The Raspberry Pi only launches a node to connect to the Arduino and the controller node. The controller node then sends commands to the Arduino through the */sailboat/sailboat_cmd* and */sailboat/sailboat_msg* topics at a regular frequency, chosen by the user.

3.3 Safety

Multiple security measures are incorporated in the Arduino and Raspberry Pi architectures. For that purpose, watchdogs are both implemented inside the Arduino and the Raspberry Pi.

To be sure that the Raspberry Pi is still on or is not freezing, it sends a message every 30 seconds to the Arduino. If the latter does not receive anything for 5 minutes, it will consider the Raspberry Pi to be malfunctioned. Such safety measures have been implemented automatically in the software architecture and the user does not need to write any further code for them.

The Arduino itself have securities in case of blockage. With a watchdog, the Arduino will set itself in a safe mode if nothing is happening. A watchdog is present in most microcontrollers. This feature is present in the

Arduino and acts as a hardware timer that has to be reset frequently. If the timer goes over a certain amount, it will restart the microcontroller. This timer is independent of the software and will run whether the software is blocked or not.

When a watchdog is triggered i.e. when the Raspberry Pi is not responding or the Arduino is blocked, the lowest-level controller, the Arduino, will set itself into Return Home control where it will come back to the point where it was turned on.

Another emergency trigger is the Radio Controller. If the radio controller is turned on, it takes control over the autonomous mode and ignores the Raspberry Pi commands. Two modes have been incorporated. One manual where both the sail and the rudder has to be controlled using the RC transmitter. Another semi-automatic where the sail uses an optimal sail angle based on the wind sensor, the user just needs to control the rudder.

4 Open Source

The main objective of this work is to create an accessible, generic and expandable platform for autonomous sailboat development. Everything detailed here is available online at `https://github.com/Plymouth-Sailboat`. We use the Github platform to share publicly all the work, including the electronic circuits and the source code. A lot of efforts were taken to write a detailed guide for new users to integrate this control box into their sailboat so as to develop their own controllers. This open-source work will always be maintained, updated and upgraded along the researches and any feedback to be received. This will give other laboratories the opportunity to build their own sailboats at a low cost and a reduced effort.

5 Tests and Experiments

We tested this control box on two sailboats : A 1m mono-hull sailboat and a 2m mono-hull sailboat as shown in Figure 5. Actuators and sensors were bought and installed along the control box inside empty sailboat hulls. Two tests were done, first in a lake near the University of Plymouth, then a second test was performed on the west coasts of France. Each of these sailboats uses the same GPS and AHRS sensors but the wind sensors are different. The small boat has only a wind direction sensor, while the other has also an anemometer. The actuators on each boat are different. Two different configuration files were created for those boats but the same code was running on both the Arduino and the Raspberry Pi on each boat. This involved changing only one file called the *config-Sailboat.h* which contains all the different parameters of the hardware such as : minimum and maximum actuator angles, minimum and maximum values sent by the different sensors, numbers of sensors and actuators, etc...

A couple of well-known algorithms were tested. On the Arduino side, as stated above, only low-level controllers were used. One which takes the rudder and sail command directly from the Raspberry Pi, and another one receiving the cap and applying an optimal sail angle and an appropriate angle for the rudder. When receiving the cap, for the sail angle δ_s and the rudder angle δ_r we have :

$$\delta_s = \frac{\pi}{2}\left(\frac{\cos(\psi)+1}{2}\right) \tag{1}$$

$$\begin{cases} \delta_r = r_{max}\sin(\theta-\bar{\theta}), & \cos(\theta-\bar{\theta}) \geq 0 \\ \delta_r = r_{max}sign(\sin(\theta-\bar{\theta})), & else \end{cases} \tag{2}$$

with ψ being the wind coming from the wind sensor (in the boat referential), r_{max} being the maximum rudder angle, θ the boat heading and $\bar{\theta}$ the cap. We consider the sail maximum angle of $\frac{\pi}{2}$. A tack strategy is also applied in case of upwind navigation, similar to the one in (Jaulin and Le Bars, 2012).

On the Raspbbery Pi side, a potential field method based on (Petres et al., 2012), a simple waypoint follower and a new position keeping algorithm were tested (Viel et al., 2018) with each sending commands at 10Hz. Some results are shown in Figure 5, where the sailboats are shown in the left two figures while the right two figures show the potential field method and way-point following control, respectively. The result figures show the GPS trajectory of the sailboats recorded through ROS in a *rosbag*. These primary test results have confirmed the feasibility and the functionality of the current hardware and software design in terms of accessibility, generality and extendability. The sailboats were able to follow the commands from the high-level controllers on the Raspberry Pi. For the potential field test, only an attractive goal point was put. The test was stopped and retrieved using the RC controller at the middle of the test in the bottom of the picture for emergency reasons.

Concerning the waypoint following controller, we put 3 different GPS coordinates to follow, shown in green in the picture. The boat was able to use a tack strategy to reach the first point at the bottom of the figure and then attain the other two GPS points. It went on to the first GPS coordinate again before we collected the boat. While the results show a sub-optimal route to get to the waypoints, it shows nonetheless the feasibility of such a control box, configured easily to apply controllers to the sailboat. After examinations of the test results, the navigation problem on the waypoint-follower was due to a mechanical fault of the sailboat, constraining a maximum rudder angle on one side.

Figure 5: Sailboats with the control box and test results.

6 Conclusion

We have demonstrated the development and the test of an open source autonomous sailing robot with emphasis on its accessibility, generality and extendability. The hardware architecture is designed to accept most common sensors and communication protocols. It is made of accessible and economic products and a customized PCB board. The software architecture is designed to comply with the hardware by being generic and extensible. The goal is to create a platform of autonomous sailing robots for generic education and research purpose. This platform can also be beneficial for other relevant researches. Future work will focus on making the whole system more robust, economic and smaller while maintaining the same level of accessibility. It will also be tested on other formats of sailing robots such as catamarans and trimarans. The research will be shared along the work progresses and more control algorithms for various autonomous missions such as area scanning and obstacle avoidance will be implemented on it in the future work.

6.0.1 Acknowledgements

This work was supported by the Engineering and Physical Sciences Research Council (EPSRC) of the U.K. under Grant EP/R005532/1.
We would also like to thank Professor Luc Jaulin from Ensta-Bretagne, France for providing testing facilities for us to test the developed sailboats in Brest.

References

Alvira, M. et al. (2013). *Small and inexpensive single-board computer for autonomous sailboat control.*

Benchoff, B. (2016). Pi 3 benchmarks: The marketing hype is true. `https://goo.gl/Nx7df2`. Accessed: 01/06/2018.

Cabrera-Gámez, J. et al. (2013). An embedded low-power control system for autonomous sailboats. In *Robotic Sailing.*

Cousins, S. (2011). Exponential growth of ros [ros topics]. *Robotics & Automation Magazine, IEEE.*

Cruz, N. A. and Alves, J. C. (2008). Autonomous sailboats: An emerging technology for ocean sampling and surveillance. In *OCEANS.*

Domínguez-Brito, A. C. et al. (2015). A-tirma g2: An oceanic autonomous sailboat. In *Robotic Sailing*.

Elkady, A. and Sobh, T. (2012). Robotics middleware: A comprehensive literature survey and attribute-based bibliography. *Journal of Robotics*.

Ghani, M. H. and Hole, a. o. (2014). The sailbuoy remotely-controlled unmanned vessel: Measurements of near surface temperature, salinity and oxygen concentration in the northern gulf of mexico. *Methods in Oceanography*, 10:104–121.

Gregory, J. (2014). *Game engine architecture*. AK Peters/CRC Press.

Jaulin, L. and Le Bars, F. (2012). A simple controller for line following of sailboats. In *Robotic Sailing 2012*.

Krauss, R. (2016). Combining raspberry pi and arduino to form a low-cost, real-time autonomous vehicle platform. In *ACC*.

Lam, T. L. et al. (2016). System design and control of a sail-based autonomous surface vehicle. In *ROBIO*, pages 1034–1039.

Larabel, M. (2018). Raspberry pi 3 benchmark. `https://goo.gl/LkDz46`. Accessed: 01/06/2018.

Lazarin, N. M. and Pantoja, C. E. (2015). A robotic-agent platform for embedding software agents using raspberry pi and arduino boards. *9th Software Agents, Environments and Applications School*.

Liu, Z. X. et al. (2016). Unmanned surface vehicles.: An overview of developments and challenges. *Annual Reviews in Control*.

Marlin. Open source 3d printer firmware. `http://marlinfw.org/`. Accessed: 01/06/2018.

Meinig, C. et al. (2015). The use of saildrones to examine spring conditions in the bering sea: Vehicle specification and mission performance. In *OCEANS 2015 - MTS/IEEE Washington*, pages 1–6.

Miller, P. et al. (2014). *MaxiMOOP: A Multi-Role, Low Cost and Small Sailing Robot Platform*.

Naveau, M. et al. (2013). Marius project: Design of a sail robot for oceanographic missions.

Neal, M. (2006). A hardware proof of concept of a sailing robot for ocean observation. *Ieee Journal of Oceanic Engineering*.

O'Hara, W. J. (2017). ASTERiaS: Autonomous Sailboat for Titan Exploration and Reconnaissance of Ligeia Sea. In *Lunar and Planetary Science Conference*, Lunar and Planetary Science Conference.

Petres, C. et al. (2012). A potential field approach for reactive navigation of autonomous sailboats. *Robotics and Autonomous Systems*.

Plumet, F. et al. (2015). Toward an autonomous sailing boat. *Ieee Journal of Oceanic Engineering*.

Rathour, S. S. et al. (2017). *Development of a Robotic Floating Buoy for Autonomously Tracking Oil Slicks Drifting on the Sea Surface (SOTAB-II): Experimental Results*.

Rules, R. (2013). Rules for roar racing. `https://goo.gl/2XjucQ`. Accessed: 01/06/2018.

Sailboat, P. (2018). Open-source source code and documentation of the plymouth sailboat project. `https://github.com/Plymouth-Sailboat`. Accessed: 01/06/2018.

Santana-Jorge, F. J. et al. (2017). A component-based c++ communication middleware for an autonomous robotic sailboat. In Øvergård, K. I., editor, *Robotic Sailing*.

Silva Junior, A. G. et al. (2016). Towards a real-time embedded system for water monitoring installed in a robotic sailboat. *Sensors (Basel)*.

Viel, C., Vautier, U., et al. (2018). Position keeping control of an autonomous sailboat. In *Proceedings of the 11th IFAC CAMS 2018*.

Wirz, J. et al. (2015). Aeolus, the eth autonomous model sailboat. In Friebe, A. and Haug, F., editors, *Robotic Sailing*.

Peruagus - a Transatlantic Autonomous Surface Vessel for the Microtransat Challenge

Elettra Ganoulis Adam Alcantara Julian Niedermaier Robert Winn
Nicholas Jones Antonio Mazzone Tur James Blake Nicholas Townsend

The University of Southampton

Abstract

The Microtransat Challenge is a (friendly) transatlantic, unmanned boat race, aimed to stimulate the development of autonomous boats. Since the first transatlantic Microtransat race in 2010 there have been over 20 entries and no successful crossings in all classes (sailing, non sailing), divisions (autonomous, unmanned) and routes (East to West, West to East). This paper presents the design and development of Peruagus, the University of Southampton 2018 Microtransat transatlantic autonomous surface vessel entry. Peruagus, meaning Globetrotter in Latin, was developed as part of a final year group design project at the University of Southampton. The design of the vessel (a mono-hull, self righting, solar powered vessel) including the system architecture, hull design, propulsion, steering, power and control systems and experimental results from a series of self propulsion tests, sea-keeping tests and autonomous operations are presented. The results demonstrate that the vessel is able to self right, propel itself with low power and operate autonomously over a range of conditions. In addition, performance predictions are presented and based on a fault tree analysis the vessel is currently predicted to have a 60% chance of success. The vessel is planned to be launched in the summer of 2018.

1 Introduction

1.1 The Microtransat Competition

The Microtransat Challenge, a transatlantic unmanned boat race (Figure 1), aims to stimulate the development of autonomous boats through friendly competition. The competition, first conceived by Mark Neal (Aberystwyth University) and Yves Briere (ISAE) in 2005, was first attempted in 2010 by Pita from Aberystwyth University (Microtransat, 2018a). Since the first transatlantic Microtransat race in 2010 there have been over 20 entries. Although the challenge is simple; autonomously travel either between Europe and the Caribbean (east to west route) or North America and Ireland (west to east route) in the fastest possible time, as of writing there have been no successful crossings in all classes (sailing/non sailing), divisions (autonomous/unmanned) and routes (East to West/West to East).

In: S. M. Schillai, N. Townsend (eds.): Proceedings of the International Robotic Sailing Conference 2018, Southampton, United Kingdom, 31-08-2018

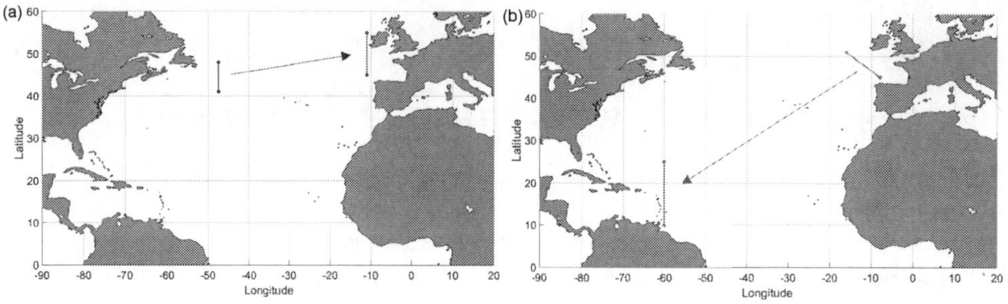

Figure 1: The Microtransat Challenge ((a) West to East route (blue), (b) East to West route (red)).

A breakdown of entries by class and division and failures is given in Table 1. The majority of entries are in the sailing class (using wind as their propulsion power) and entered in the autonomous division. Reviewing the known failures, the technical failures primarily relate to issues of reliability - surviving in the harsh environment for a prolonged period of time. While the non-technical failures relate to route hazards - fishing grounds, shipping lanes and the Sargasso Sea (an ocean gyre off the coast of northern America and the Caribbean characterised by brown seaweed, which creates obstacles for the vessel).

	Sailing Class Only wind power can be used for propulsion, overall length (LOA) restricted to a maximum of 2.4m	**Non-Sailing class** Any type of propulsion can be used, overall length (LOA) restricted to a maximum of 2.4m
Autonomous (Division) No interaction between the team and the vessel, only publicly available data can be received by the vessel (i.e. no waypoint changes	Pinta, Snoopy Sloop 10, Snoopy Sloop 11, Breizh Tigress, Opentransat *Erwan 1, Aboat Time, Trawler Bait, Phil's Boat*, Breizh Spirit, Breizh Spirit DCNS, Snoopy Sloop 8, Snoopy Sloop 9, That'll do	That'll do two (Epsom College Entry)
Unmanned (Division) Data can be sent to the boat, including course changes	Gortobot V2, *SB-wave*	

Table 1: Summary of Microtransat classes, divisions and failures by vessel name. (*Italics denotes a non-technical failure e.g., picked up by fishing vessel*, Underline denotes technical failure e.g., position report failure, unmarked denotes unknown or sailed into land). Data from (Microtransat, 2018b)

Furthermore, considering vessel size and performance (time sailed and distance covered) there are no apparent trends, Figure 2. Neither is there a clear improvement in performance over the years the competition has been running, although this can be attributed to the small dataset and difficulty of the challenge. In this regard it is hoped this paper will provide a valuable insight for new teams and entires in the Microtransat.

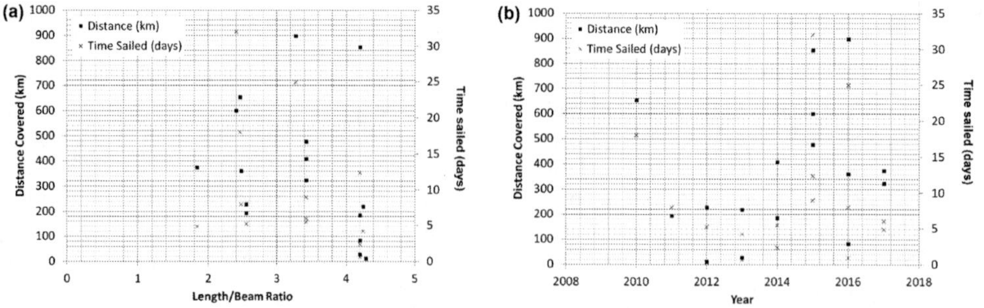

Figure 2: Comparisons of previous Mircotransat Entries ((a) Length/beam ratio versus distance and time sailed (b) Year versus distance and time sailed)

1.2 The Peruagus Project

Peruagus, meaning Globetrotter in Latin, is the University of Southampton 2018 Microtransat transatlantic autonomous surface vessel entry. The Peruagus project was a final year engineering group design project at the University of Southampton. The group design projects (GDPs) at the University of Southampton (University of Southampton, 2018) aim to provide students with the opportunity to demonstrate their knowledge and skills, gained during their degree, to a 'grand' engineering design challenge. In this regard competitions can be used to great effect, as reported by (Telegraph, 2016) and exemplified by Xprize (Xprize, 2018), Eurobot (Eurobot, 2018), formula student (IMechE, 2018) and maritime engineering related competitions including the World Robotic Sailing Championships (WRSC, 2018) and the International HydroContest(Hydros Foundation, 2018). In particular, the Microtransat competition has provided a multi-faceted, challenging, motivational, open-ended engineering problem. This has enabled students to demonstrate and integrate knowledge acquired from across their programs but also provided the opportunity to interact and contribute to an international community.

The aim of the Peruagus project is to design and develop a vessel to cross the Atlantic, as part of the Microtransat Challenge. Since the vessel is required to operate unmanned and travel for several months autonomously without maintenance and with all previous attempts unsuccessful, a failure analysis approach was used to guide the design of Peruagus, focusing on reliability (minimising the probability of system and subsystem failures to maximise the chances of success).

1.3 Contribution and Paper Structure

In this paper, the design of Peruagus, a mono-hull, self righting, solar powered vessel, is presented. The system architecture, hull design, propulsion, steering, power and control systems are detailed in Section 2, including experimental results from a series of self propulsion tests, sea-keeping tests and autonomous operations. The final vessel design, targeting the Mircotransat non-sailing class, autonomous division (with the possibility to convert to unmanned) following an east-west route, is presented in Section 3 and performance predictions are presented in Section 4.

2 Peruagus Vessel Design

2.1 System Architecture

An overview of Peruagus system architecture is given in Figure 3. The electrical system is split into a pair of redundant circuits, each comprising of a 100W solar panel made up of SunPower cells, a Victron BlueSolar MPPT charge controller, a battery bank (2 x 480Wh lithium-ion batteries), control relays and a step down voltage converter. The solar/battery bank (12-14V) provides power to the drive motors, steering actuators, the navigation light and the bilge pump. A step-down voltage converter (also powered from the solar/battery bank) then provides a regulated 5.6V supply for the on-board control systems, including;

1. The navigation controller (Pixhawk), which acts as the navigation controller, and manages the power distribution and makes decisions based upon the condition monitoring sensors
2. The satellite modem (RockBLOCK+) which transmits telemetry and location data every 6 hours using the Iridium network
3. The microcontroller (Teensy 3.5), which provides data acquisition, processing, logging and (I2C) communication with the navigation controller.

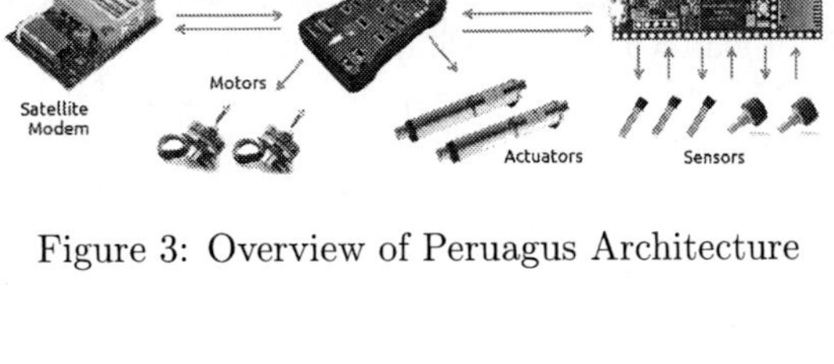

Figure 3: Overview of Peruagus Architecture

2.2 Hull Design

To house all the systems a mono-hull, self righting, solar powered vessel with passive keel cooling was developed. The hullform, as shown in Figure 4, was made of a double skinned foam core (Celotex PIR wall insulation foam, milled from a foam block using a CNC machine) with E-glass ($290g/m^2$ with $100g/m^2$ finish) infused with EL2 epoxy resin. The foam thickness has a total volume of approximately 100kg displacement. This ensures that if there is water ingress and the compartment becomes flooded the vessel will maintain positive buoyancy. The propeller shafts were also angled at 15 degrees, to ensure the stern tubes ends were above the waterline such that in the event of a seal failure water would not flood the boat.

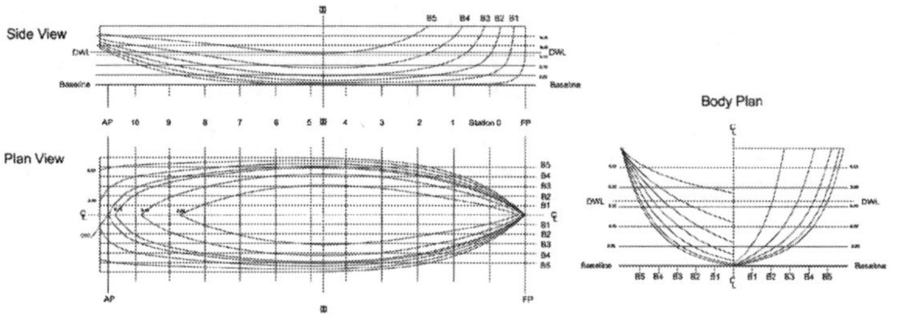

Figure 4: Peruagus hullform

2.2.1 Stability

Given the slender form, solar panel requirements and potentially severe sea-states, Peruagus was designed to be self righting and remain self righting in the event of damage and flooding. This was achieved with a keel, watertight compartments and an asymmetric superstructure. The keel (NACA0010 section) was made from PET plastic, housing 16kg of lead ingots constructed around an aluminum frame. The GZ curves for the intact vessel and damaged vessel are given in Figure 5.

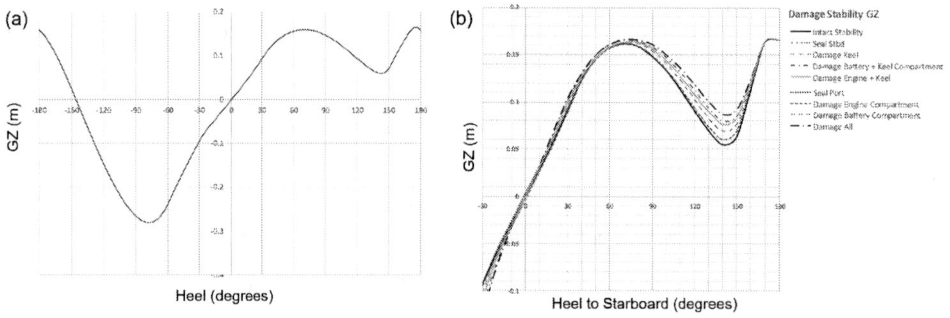

Figure 5: Peruagus GZ stability curves ((a) Intact (b) Damaged)

2.2.2 Seakeeping

To assess the seakeeping performance of Peruagus a series of observational experiments were performed in the University of Southampton towing tank, including a worse case scenario when the wave length was equal to the LBP, Figure 6. As the vessel is unmanned and the limiting factor for seakeeping performance is the tolerance of the electronics, this approach which does not necessitate the testing of every possible sea-state that may be encountered, significantly reduced the number of tests required. The results, Figure 6, show that the accelerations experienced by the vessel are within tolerance of all electrical systems. For example, the most sensitive system onboard, the GPS, is rated to withstand up to 4G acceleration. While the wave height is limited in the towing tank, these tests provided confidence in the seakeeping performance of the vessel.

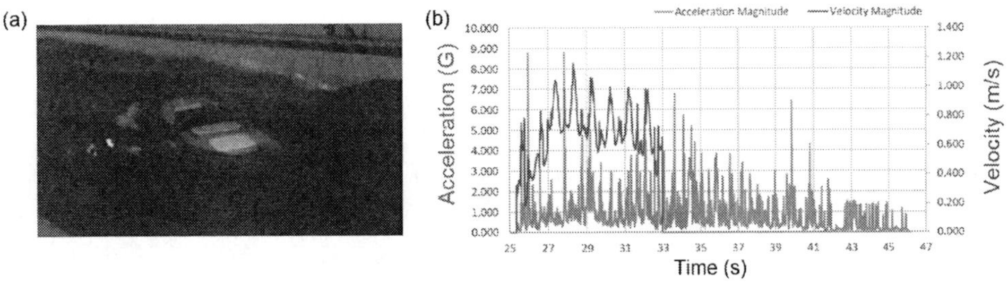

Figure 6: Peruagus Seakeeping ((a) Image of test, (b) Heave acceleration and velocity recorded using an IMU)

2.3 Propulsion System

Propulsion was achieved with two inwards rotating propellers, providing (if necessary) differential thrust for steering in the event of a rudder malfunction. To establish the resistance and estimate the required propulsion power for the hull, the viscous resistance was estimated using the ITTC 1957 correlation line (Molland et al., 2017) with a form factor identified from CFD simulations (using ANSYS Fluent) and the wave resistance coefficients were determined using Maxsurf resistance Slender Body Analysis (assuming Peruagus can be regarded as a fully displacement traditionally shaped vessel). The results are shown in Figure 7. The power estimates were made assuming the following efficiencies; Propeller Angle Efficiency 96.59%, Propeller Efficiency 60%, Transmission Efficiency 95% and a Weather and Fouling Margin 30%. Based on the results, a 140W design specification was considered (enabling a nominal 45-55W 'cruise' operation and 'sprint' ability to maintain progress in adverse conditions or in the event of an engine, belt, shaft or propeller failure).

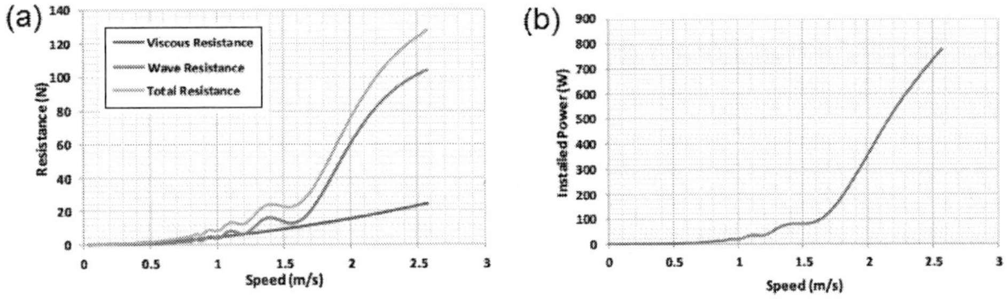

Figure 7: Estimated resistance (a) and power (b) over a range of speeds

2.3.1 Drive System

A pair of brushless hobby-grade (Turnigy DST-700kv, 12V) motors were selected for propulsion. With a nominal no load speed of 8,400rpm (12V), a 1:4.4 geared belt drive system was implemented to provide the propeller design speed of 650-800rpm. To check the longevity of the motor and controller, the motor was run for over 3000 hours, with an applied load (an airscrew), cycling between cruise (20-40% throttle) and sprint throttle levels to provide a representative load cycle. Although, an increase in bearing noise was noted, the motor and controller ran without fault with no notable increase in power consumption, providing confidence in the selected drive system.

2.3.2 Propeller Selection

The propeller selection was based on an experimental investigation of 4, 5 and 6 bladed Wageningen B-series propellers (Van Lammeren et al., 1969) (readily available brass model boat propellers designed to operate at 750 rpm at 1.5knots). The 4, 5 and 6 bladed propellers were 3D printed (for the experiments) in high density ABS and vessel speed, power (current drawn) and rpm, over a range of throttle settings, in the University of Southampton Boldrewood towing tank, were recorded, Figure 8. Based on the results, Figure 9, the five bladed propeller was selected. The results show a slight discrepancy between the theoretical estimates (see section 2.3) where an installed power requirement of 8.07W at 1.5knots 'cruise' speed (0.77m/s, Fn0.168, 4.43N resistance,

4.58N thrust), and installed power requirement of 80.8W at 3.0knots 'sprint' speed (1.54m/s, Fn0.336, 22.18N resistance, 22.96N thrust) was calculated.

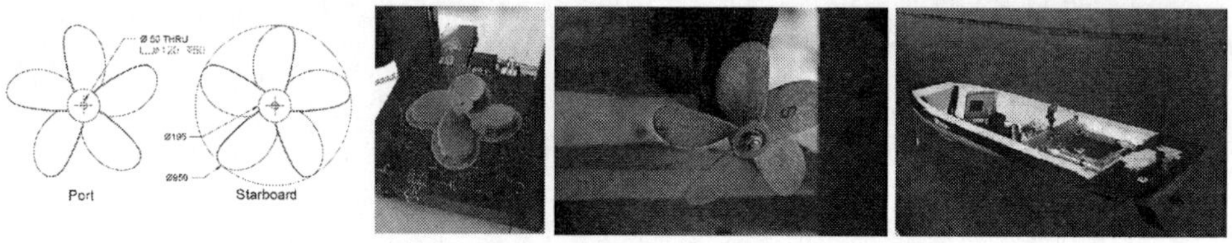

Figure 8: 3D printed propeller (design, manufacture, assembly, testing)

According to the results of the self-propulsion tests, the motors operate at approximately 40% (each) at normal cruise speed, drawing a total of 48W. One motor was found to push the vessel at a maximum speed of 2 knots at full applied power (drawing approximately 80W). While both motors operating at full applied power, produced a speed of just below 3 knots (drawing 160W total). Based on the results the boat is intended to be operated for the majority of journey at 'cruise' power (45-55W continuous input), with the ability to 'Sprint' to maintain progress when encountering tidal streams, current, or heavy weather.

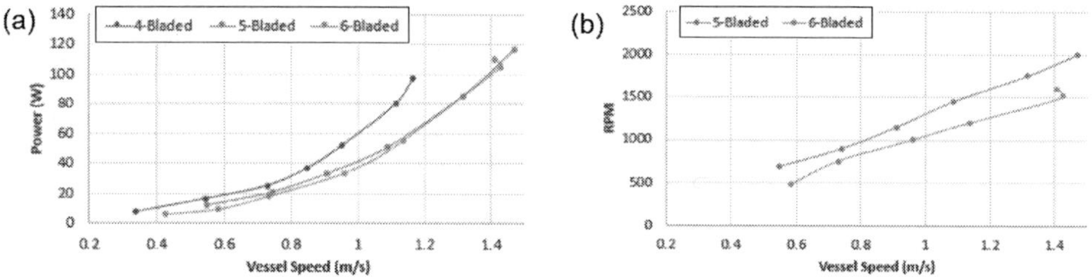

Figure 9: Propeller test results ((a) Power (b) RPM)

2.4 Steering System

To steer the vessel a doubly redundant system was developed with two linearly actuated control surfaces (rudders) and two inwards rotating propellers (if necessary in the event of rudder failure) providing differential thrust. The rudders were designed to manoeuvre the vessel and, in case of a propulsion failure, counteract the moment produced by a single, one-sided propeller. This approach was adopted, based on advice provided by ASV Global Ltd and on the main causes of failure of small ocean going craft; failed rudder servos (e.g., from salt water seizing the electronics or strong forces damaging the actuator) (Microtransat, 2018b) and complete rudder loss (Seacharger, 2018). In addition, linear actuators also have the added advantage that they can hold their position without the need for power.

2.5 System Power

To power Peruagus an asymmetric superstructure solar battery charging system was implemented. Since the transat is in the northern hemisphere and one way - east to west, the asymmetric superstructure maximises the incident solar energy and additionally aids in self righting (with a tendency to right itself to starboard). In total two 100W (18V) Mono-crystalline solar panels (η = 23.5%), were selected and wired in parallel, with each panel charging two Lithium Ion batteries (12V, 3S 40Ah) (with a Victron solar controller and built-in profile for Lithium Ion cells; lower voltage cutoff of $\approx$ 8.5V and a maximum voltage of 12.6V).

2.6 System Control

The control system is based on the Ardupilot Rover, using GPS to steer the vessel between waypoints as shown in Figure 11. A major modification to the Ardupilot Rover basis firmware is the addition of the Director, which acts as a proxy between the autopilot and the physical hardware. The autopilot code sends a desired throttle and steering value to the Director. The Director, which continually monitors the vessel to detect failures, then drives the motors and rudders accordingly (given the system status) following prescribed rules. In addition, the Director also monitors longer term navigational performance e.g., IMU data to detect the severity of boat motions, ability to maintain headway, capsize events and system failures. This data can be reported to the shore base and actions including; adding/changing waypoints, overriding autopilot decisions on failure, change the data sent in the status message, request a full diagnostic to be sent (multiple messages), manually set throttle or steering angles, switch to drift/loiter mode and conduct a reverse 360, can be taken. Meaning that, in the event of an unforeseen failure there is the possibility to convert to the unmanned division. Although, the 'rules' or thresholds remain to be finalised, initial tests have been conducted to provide confidence in the control, with Peruagus run autonomously for an hour tracking between waypoints in a local lake.

3 Final Peruagus Design

The final Peruagus design is presented in Figure 10 and Table 2.

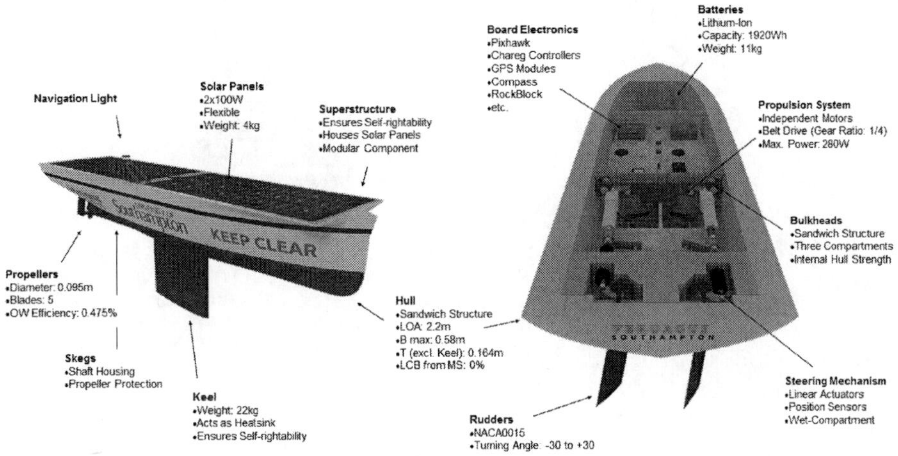

Figure 10: Peruagus General Arrangment

HULL PARAMETERS	
Length (LOA, LWL)	2.2m, 2.158m
Beam (WL, Max)	0.475m, 0.58m
Draft at FP (at AP)	0.162m (0.14m)
Displacement	80.4kg
Block Coefficient, CB	0.454
LCB, LCF, GMt, Trim	1.16m, 1.071m, 0.167m, -0.021m
PROPULSION SYSTEM	
Propellers	2× 5-bladed Raboesch M5 propellers, 110mm diameter
Propeller drive	2× Model DST-700, 140W, 700RPM/volt, brushless
STEERING SYSTEM	
Rudder Shape	NACA0015 (widened near the stock)
Rudder Material	Acrylonitrile Butadiene Styrene (ABS) and Epoxy
Rudder Area	$0.014m^2$
Mean Chord, Span, Sweep	0.10m, 0.14m, 20°
Aspect Ratio	1.4
SYSTEM POWER	
Solar Panels	2× 100W (18V), Monocrystalline, 1050mm x 540mm x 2.5mm, $\eta = 23.5\%$
Batteries	4× 480WH, 12V (Nominal) Lithium Ion (3S 40Ah), rated discharge 40A, rated charge 10A, 12.50kg (total)
CONTROL SYSTEM	
Navigation Controller	Pixhawk
Satellite Modem	RockBLOCK
Microcontroller	Teensy 3.5

Table 2: Peruagus particulars

4 The Peruagus Entry

4.1 Vessel Route

Peruagus is planned to be entered into the Mircotransat non-sailing class, autonomous division (with the possibility to convert to unmanned in the event of a required interaction) following an east-west route. To minimise the probability of failure the planned route (as much as practically possible) avoids fishing grounds, shipping lanes and the Sargasso Sea, Figure 11. Although, the east-west route is longer and arguably more challenging, with a 91% probability that one hurricane will be encountered in the course of the transit (NOAA, 2018), it is practical for a UK team and the average wind and waves directions are favourable.

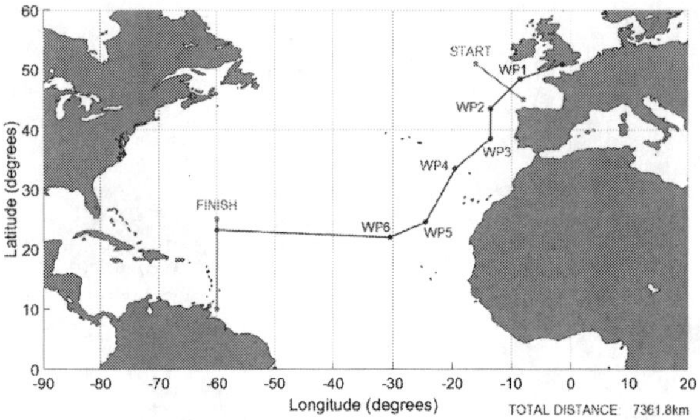

Figure 11: Peruagus Route

4.2 Estimated Duration

Using the Haversine formula (Mwemezi and Huang, 2011) to calculate the distances between the planned waypoints (latitude,longitude) and assuming a constant 'cruise' speed of 1.5knots (55W), the total journey is expected to take around 16 weeks, Table 3. Although neglecting the influence of any current, wind and waves, these estimates will enable a comparison with the actual performance and potentially highlight performance issues on route.

Waypoints	Latitude (N)	Longitude (W)	Distance (km)	Propulsion Energy (MJ)	Duration (Weeks)
Southampton	50.91°	1.40°			
Waypoint 1	48.50°	8.50°	565.24	40.3746	1.2138
Waypoint 2	43.50°	13.50°	667.05	47.6461	1.4324
Waypoint 3	38.50°	13.50°	555.97	39.7125	1.1939
Waypoint 4	33.50°	19.50°	762.88	54.4911	1.6381
Waypoint 5	24.50°	24.50°	1103.31	78.8080	2.3692
Waypoint 6	22.00°	30.50°	667.76	47.6972	1.4339
Finish line	23.20°	60.00°	3039.62	217.1155	6.5270
		TOTALS	7361.83	525.8451	15.8082

Table 3: Peruagus estimates (assuming a constant cruise speed of 1.5knots at 55W)

4.3 Probability of Success

To determine the probability of success a deductive failure analysis in the form of a fault tree was conducted, as illustrated in Figure 12. Over 200 identified events (each representing a possible cause of failure of one of the vessels systems) were identified and probabilities of failure assigned. The probabilities were based on the test results, available literature, however, some probabilities were difficult to quantify and estimates were made. The final analysis estimated the probability of success at approximately 60%.

Given the history of the competition this figure seems reasonable, however it is important to note that there is a degree of uncertainty associated with this number. For example, the largest contribution to vessel failure (18%) is attributed to the inability of the Pixhawk (and the vessels operating code) to handle

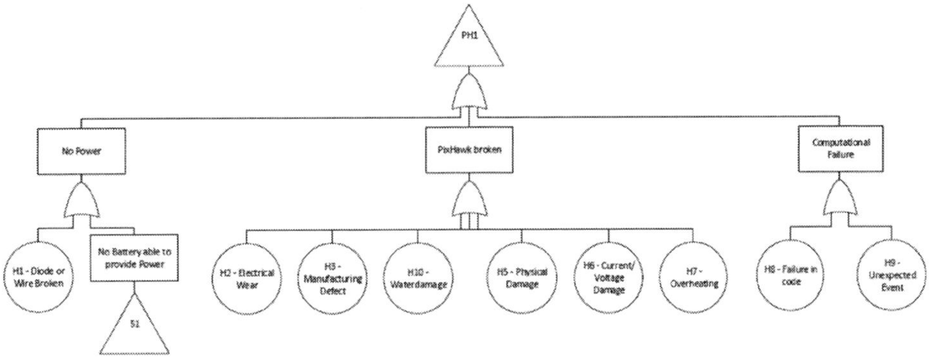

Figure 12: Example branch of the fault tree analysis

an event not part of the vessels normal operation and the lack of code specifically designed to handle the event. Since the probability of such an event is unknown, this is a subjective estimate by the team. Although subjective, this approach does provide a means to quantify the probability of success and, more usefully, to enable designs to be compared and developed with a quantifiable metric to maximize the probability of success.

5 Conclusion

This paper presented the design of the Peruagus, the University of Southampton 2018 Microtransat transatlantic autonomous surface vessel entry. The final vessel design is presented, including the system architecture, hull design, propulsion, steering, power and control systems. Experimental results are presented demonstrating that the vessel is able to self right, propel itself with low power and operate autonomously over a range of conditions. Furthermore, performance predictions are presented and based on a fault tree analysis the vessel is currently predicted to have a 60% chance of success. The vessel, a mono-hull, self righting, solar powered vessel is planned to be launched in 2018, in the Mircotransat non-sailing class, autonomous division, following an east-west route.

Acknowledgements

The support of BMT Argoss and ASV Unmanned Marine Systems is acknowledged in the development of this project, in addition to Bertrand Malas, towing tank manager, Andy Robinson in TSRL and Dave and Terry of the student workshop.

References

Eurobot (2018). Eurobot international robotic contest. http://www.eurobot.org/. Last called on 30th July 2018.

Hydros Foundation (2018). Hydrocontest. ttp://www.hydrocontest.org/en/event.html. Last called on 30th July 2018.

IMechE (2018). About formula student. http://www.imeche.org/events/formula-student/about-formula-student. Last called on 30th July 2018.

Microtransat (2018a). The microtransat challenge. https://www.microtransat.org/index.php. Last called on 30th July 2018.

Microtransat (2018b). The microtransat challenge history. https://www.microtransat.org/history.php. Last called on 30th July 2018.

Molland, A. F., Turnock, S. R., and Hudson, D. A. (2017). *Ship resistance and propulsion.* Cambridge university press.

Mwemezi, J. J. and Huang, Y. (2011). Optimal facility location on spherical surfaces: algorithm and application. *New York Science Journal*, 4(7):21–28.

NOAA (2018). National hurricane center data archive. https://www.nhc.noaa.gov/data/. Last called on 30th July 2018.

Seacharger (2018). Seacharger oceangoing autonomous boat. http://www.seacharger.com/. Last called on 30th July 2018.

Telegraph (2016). Competitions help students gain real-life skills. https://www.telegraph.co.uk/education/stem-awards/energy/competitions-help-students/. Last called on 30th July 2018.

University of Southampton (2018). Uos design show 2018. http://uosdesign.org/designshow2018. Last called on 30th July 2018.

Van Lammeren, W., Van Manen, J., and Oosterveld, M. (1969). The wageningen b-screw series.

WRSC (2018). World robotic sailing championship. https://www.roboticsailing.org/. Last called on 30th July 2018.

Xprize (2018). Xprize homepage. https://www.xprize.org/. Last called on 30th July 2018.

Dynamic Simulation Model for an Autonomous Sailboat

Moritz C. Buehler Carsten Heinz Simon Kohaut

Sailing Team Darmstadt e.V.
{moritz.buehler, carsten.heinz, simon.kohaut}@sailingteam.tu-darmstadt.de

Abstract

Sailing without any human intervention generates a great fascination, since it is challenging while enabling many opportunities. Moving without external energy supply, possibly transporting goods, collecting plastic waste or recording scientific measurement data, new attractive scenarios become possible. As first milestone, we aim to send a robotic sailboat across the Atlantic ocean, coping with bad weather including storms, other vessels or floating waste.

For testing, designing control, sophisticated route planning, and managing algorithms, we are interested in differential equations for a simulation model. Therefore, we theoretically derive relations, describing the sailboat motion as those of a rigid body having six degrees of freedom (6 DOF). This allows us to reproduce significant nonlinear effects like the ones created by flow separation, speed dependence of the dynamics and oscillations by waves, while maintaining a comprehensible structure of the external forces of the sailboat. In contrast to standard velocity prediction programs (VPP), we are interested in the actual dynamical behavior, the effect of sail angles, rudder and waves while the precise prediction of the actual reachable velocities are of minor interest.

The dynamic model can serve for controller design and for the generation of test data. Additionally, it is used to feed a hardware model of our boat, providing an intuitive way of demonstrating the boat movements.

1 Introduction

For large parts of the oceans, accurate measurements of the environment are missing. More detailed data would benefit climate research and environment monitoring. A way to perform the ocean sensing is the usage of autonomous sailing boats. This reduces the cost of the required hardware from a big research vessel to a small boat and eliminates personnel costs as there is no crew on the ship. Further use cases are the collection of trash and cheap, energy efficient transportation of goods.

These long term missions through rough weather conditions require a high confidence in the hardware of the boat and in the algorithms. Therefore, a way to test the implemented controller is required. In this paper, we present a simulator to provide a basis for testing our developed algorithms and systems of an autonomous

In: S. M. Schillai, N. Townsend (eds.): Proceedings of the International Robotic Sailing Conference 2018, Southampton, United Kingdom, 31-08-2018

sailboat. This allows a short development cycle with fast feedback without the need to use a real boat. The focus is on a six degrees of freedom (6 DOF) model for a sailing boat. It is used for simulating the dynamic boat movement. Further it is important for controller design and a test pattern generator for the whole software architecture as well as for visualization purposes.

1.1 Related work

For modeling the dynamics of motorized vessels there is rich literature, e.g. (Fossen, 2005), however for wind propulsion, special effects have to be considered. The specifics of sailing are treated in models for yachts (Philpott et al., 1993) in the development of Velocity Prediction Programs (VPP). However, they aim to optimize the race performance and therefore focus on the achievable speed, while we are interested in the dynamic motion effects. Dynamic models for sailing vessels are derived in (Saoud et al., 2013), (Alves, 2010), (Masuyama and Fukasawa, 2011) and (Roncin and Kobus, 2004). However, (Masuyama and Fukasawa, 2011) concentrates on tacking, (Saoud et al., 2013) and (Alves, 2010) reduce the model to 3 DOF, neglecting the influence of roll and waves. (Roncin and Kobus, 2004) relies on the identification of a larger set of parameters based on experiments for which we would need a setup respectively CFD model. In contrast, we outline a 6 DOF model including wave effects while transparently formulating the acting forces using fewer parameters to obtain a rich but comprehensible simulation setting.

1.2 Assumptions

For our model, we take several assumptions. First, we neglect the change of the wind over the height and assume a constant wind speed as well as angle over the whole sail. A similar assumption is made for the lateral parts and the water speed. Also the interference between both sails is neglected. Each sail will change the wind locally which will have effects on the other sails. Instead, lift and drag coefficients of the sails are approximated by the results of thin airfoil theory, which additionally neglects the shape change of the sail. The flow separation is estimated in a simple model interpolating between unseparated and fully separated flow. For the waves, it is assumed, that the wave length is large compared to the ship length.

2 Mathematical model of the boat movement

To simulate the movement of the sailboat, we need a model in the form of differential equations, describing the temporal relations of external forces acting on the sailboat through dynamic and kinematic relations. For our use case, we are interested in a rich but comprehensible model, covering important effects of the sailboat's motion.

We model the dynamics and kinematics for a 6 DOF boat model, that depend on the geometries and inertia. A sailboat operates at the boundary between two media, water and air and uses speed differences to move forward. As important forces, we derive equations for the following components:

- sail, keel and rudder force
- buoyancy forces including effects from ocean waves
- wave resistance
- damping forces

2.1 Dynamics and kinematics

The dynamics describe how the forces acts on the boat speeds and turn rates due to the boat inertia. It is wise to derive the relations of forces and dynamics in different reference frames, that are connected through kinematic relations, see also Figure 1.

We describe the position x_g, y_g, z_g and heading ψ of the boat in our globally fixed or navigational frame, with $z = 0$ at the undisturbed water surface.

At the center of gravity of the boat, we locate the second, the heading frame, also parallel to the water surface with x axis in heading direction. The transformation from the global frame consists of the translation of position (x_g, y_g, z_g) and rotation according to heading ψ. In this heading frame, we formulate the buoyancy F_{hs}, the translation dynamics (change of speeds v_x, v_y, v_z) and the wave resistance F_{wr}. The basis vectors in the heading frame are denoted by $\{\mathbf{e}_x, \mathbf{e}_y, \mathbf{e}_z\}$.

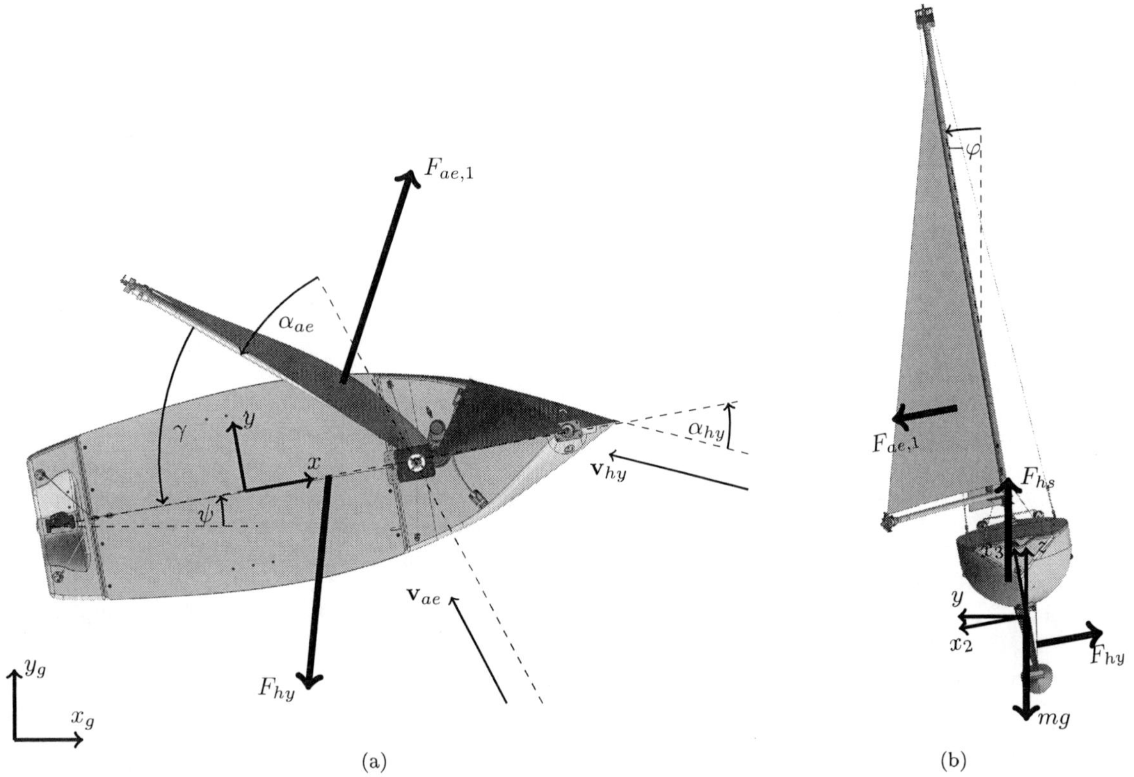

Figure 1: Top view (a) and rear view (b): Forces of sail(s) F_{ae}, keel F_{hy} and buoyancy F_{hs}. Position (x_g, y_g, z) and heading ψ is measured in a global frame, buoyancy and lateral dynamics in a frame centered at the ship but oriented to the water surface, and the angular dynamics in a ship oriented coordinate frame. The aerodynamic angle of attack α_{ae} results from sail angle γ and apparent wind direction $\beta_{WA} = \gamma + \alpha_{ae}$, hydrodynamic α_{hy} from the leeway drift and water stream.

The third frame is the body frame, fixed to the boat main axes. Here, it is intuitive to describe the rotational dynamics, sail, rudder and keel forces. The transformation from the heading frame is achieved by rotating due to roll angle φ, while the pitch angle θ is often negligible small. The basis vectors of this frame, we denote as $\{\mathbf{e}_1, \mathbf{e}_2, \mathbf{e}_3\}$.

Altogether, we have 12 states, the translations x_g, y_g, z_g, heading ψ, roll φ and pitch θ and the change rates in positions v_x, v_y, v_z and angles p, q, r, where $\mathbf{v} = v_x\mathbf{e}_x + v_y\mathbf{e}_y + v_z\mathbf{e}_z$ is the speed at center of gravity and $\Omega = p\mathbf{e}_1 + q\mathbf{e}_2 + r\mathbf{e}_3$ the angular velocity.

From the kinematics, we obtain the first part of differential equations, the position (ground based) changes according to heading and speeds,

$$\frac{d}{dt}\begin{pmatrix} x_g \\ y_g \\ z \end{pmatrix} = \begin{pmatrix} v_x\cos(\psi) - v_y\sin(\psi) \\ v_x\sin(\psi) + v_y\cos(\psi) \\ v_z \end{pmatrix}. \tag{1}$$

Similar holds for the angles, where it is used, that the pitch angle is small.

$$\frac{d}{dt}\begin{pmatrix} \varphi \\ \theta \\ \psi \end{pmatrix} \approx \begin{pmatrix} p \\ \cos(\varphi)q - \sin(\varphi)r \\ \cos(\varphi)r + \sin(\varphi)q \end{pmatrix} \tag{2}$$

The speed and angular rates change due to forces and moments according to the dynamic equations. The boat speed change due to overall applied forces $\mathbf{F}$ and boat mass m,

$$m\frac{d}{dt}\mathbf{v} = \mathbf{F}. \tag{3}$$

For the left side we need to respect the angular velocity of the heading frame, the change of heading, $m\frac{d}{dt}\mathbf{v} = m((\dot{v}_x - \dot{\psi}v_y)\mathbf{e}_x + (\dot{v}_y + \dot{\psi}v_x)\mathbf{e_y} + \dot{v}_z\mathbf{e}_z)$.

When the boat accelerates, surrounding water is accelerated, resulting in a transient hydrodynamic force that is usually modeled as an added mass. Additionally, this effect introduces cross terms, coupling roll and sway. We plan to include added masses soon in our description, since they have an important influence, especially for roll motion (Korotkin, 2009). Therefore, (3) and (4) have to be slightly adapted to include the coupling of inertia.

For the angular velocities, we have the relation of the angular momentum $\Theta\boldsymbol{\Omega}$ and overall torque $\mathbf{T}$,

$$\frac{d}{dt}(\Theta\boldsymbol{\Omega}) = \mathbf{T} = \sum_i \mathbf{r}_i \times \mathbf{F}_i, \tag{4}$$

where Θ is the angular inertia, $\boldsymbol{\Omega}$ the angular rate vector and $\mathbf{r}_i$ the vector from center of gravity to the ith force point of application. Due to approximate symmetries of the boat, the principal inertia axes do coincide with the boat main axes, leading to a diagonal inertia $\boldsymbol{\Theta}$ (apart from added masses). For the derivative, it is important to respect, that the ship itself is moving and the ship based coordinate frame itself has to be differentiated, $\frac{d}{dt}(\Theta\boldsymbol{\Omega}) = \dot{p}\Theta_1\mathbf{e}_1 + \dot{q}\Theta_2\mathbf{e}_2 + \dot{r}\Theta_3\mathbf{e}_3 + \boldsymbol{\Omega} \times \Theta\boldsymbol{\Omega}$. The formulation in ship based coordinates is however necessary to avoid a full and time dependent inertia matrix.

In the following we derive equations for the force components describing the motion of the sailboat, needed for equations (3) and (4).

For the aero and hydrodynamics, we have to differ apparent wind (stream) and the true wind (stream). The true wind is the wind referred to the ground, that is equal for all vessels at a given position. For the hydro- and aerodynamic forces, the relative flow is relevant, meaning that the ship speed has to be subtracted from the absolute. Hence the apparent or relative wind (and water flow) is calculated as the vectorial sum of the true wind (water stream) and the negative boat velocity. With raising boat velocity, the apparent wind changes its directions towards coming from ahead with varying strength. Additionally, the rotation of the boat changes the relative flow on the foils. Therefore, we chose the point of load $\mathbf{r}_{foil}$ of the foil and assume constant flow speed over the whole foil (which seems to be fine for not too large rotation rate). The relative velocity is calculated as $\mathbf{v}_{rel} = \mathbf{v}_{flow} - \mathbf{v} - \Omega \times \mathbf{r}_{foil}$. For the rudder this might be problematic due to the nearby keel but this remains for further refinement.

2.2 Sails and lateral foils

A main impact results from the sails and lateral plans, keel and rudder. Both, we model analog as ideal foils in a fluid stream. We calculate the resulting fluid forces based on results from theoretical aerodynamics, providing equations for lift and drag forces and the points of attack.

In difference to a hard foil where the actual shape is fixed, the shape of a sail varies according to load and design. However, we want to stay at a simple description combined with fast computation and neglect these effects.

2.2.1 Lift and drag

Lateral foils and sails are placed in a fluid stream that reaches them under a specific angle of attack α. For the keel, this is the relative angle of the water flow. For the sails, the apparent wind angle is referred to the sail angle γ (see Figure 1).

For the forces, we use the results from theoretical aerodynamics of thin foils leading to the relation for the lift coefficient as

$$c_L = c_{L,0} + c_{L,\alpha}\alpha, \tag{5}$$

where $c_{L,0}$ is the lift coefficient at zero angle of attack that is zero for a symmetric foil, and for thin foils a theoretic value of the slope coefficient of $c_{L,\alpha} = 2\pi$ is known (while neglecting the foil shape) (Spurk and Aksel, 2010).

The drag force of a foil consists of different components: Friction at the surface, induced drag due to finite foil height and pressure loss when it comes to flow separation that we treat later. At small angles of attack, the drag coefficient is

$$c_D = c_f + c_{D,i}. \tag{6}$$

The induced drag depends quadratically on the lift coefficient $c_{D,i} = c_L^2/(\pi\Lambda)$ with the geometric foil stretching Λ, height h and area A. The theoretical friction coefficient of a plate in laminar flow is $c_f = 2.66/\sqrt{Re_l}$ (Spurk and Aksel, 2010), where $Re_l = \frac{vl}{\nu}$ is the Reynolds number with length l of the plate in direction of the flow and ν is the kinematic fluid viscosity.

With drag and lift coefficients we can calculate the forces by multiplying the dynamic pressure $q = 0.5\rho v_{rel}^2$ and the area A,

$$\mathbf{F}_{flow} = qA\left[(\sin(\beta_{WA})c_L - \cos(\beta_{WA})c_D)\mathbf{e}_1 - (\sin(\beta_{WA})c_D + \cos(\beta_{WA})c_L)\mathbf{e}_2\right], \tag{7}$$

that is transformed to the ship based frame according to the relative flow angle β_{WA}.

The point where the aerodynamic force acts, is for the flat symmetric wing at one forth length, but varies until one third for asymmetric shapes where the location depends on the angle of attack (Spurk and Aksel, 2010).

2.2.2 Flow separation

At larger angles of attack (e.g. sailing in front of the wind for the sail, departing from low speeds for the keel), the flow separates from the foil, the lift reduces and the drag by the pressure loss gains significant influence. In the extreme case of an angle of attack of 90 degrees, there is no lift (due to symmetries), while the drag coefficient is slightly above $c_D \approx 1$ and the point of attack is the center of the foil. At smaller angles, we model the drag coefficient as $c_D = \sin(\alpha)^2$ representing the dynamic pressure of the flow perpendicular to the foil. For most foil shapes, flow separation leads to stall angles, the angle where c_L is maximal, between 15 and 25 degree. We interpolate between not separated case and flow separation heuristically according to a separation factor $s = 1 - \exp\left[-(\alpha/\alpha_{sep})^2\right]$ and use a characteristic angle $\alpha_{sep} = 25°$, resulting in a stall angle of 15°.

2.3 Wave resistance

The velocity of a displacement vessel is limited by the wave resistance at maximum hull speed. This resistance results from the waves generated by the ship movement. At about a Froude number of $Fr = v/\sqrt{gl_{wl}} = 0.4$, the wave resistance reaches a local maximum that cannot be exceeded by our keel boat.

The maximum hull speed depends on the square root of the waterline length l_{wl} of the ship $v_{hull} = 0.4\sqrt{gl_{wl}}$, resulting to a maximum speed for a 4 m water line length of about 2.5 m/s.

We approximate the wave drag by a 6th order polynomial introducing it as speed limiting factor, neglecting the hull shape dependent interference effects and model the wave resistance according to

$$\mathbf{F}_{wr} = -\operatorname{sign}(v_x)c_{wr}q_{hy}A_{LK}\left(\frac{v_{hy}}{v_{hull}}\right)^4\mathbf{e}_1, \tag{8}$$

with lateral area A_{LK} and a wave resistance weight parameter c_{wr}.

2.4 Hydrostatic buoyancy and wave influence

The hydrostatic force is central for the boat to float on the water and the moment stabilizes roll and pitch angles.

The hydrostatic bouyancy respectively Froude-Krylov force (Fossen, 2005) is the integrated water pressure over the hull, also described as the weight of displaced water. For the calculation of the pressure, the virtual water surface without the vessel has to be used. In calm water and zero speed, the hydrostatic force acts vertically (along z_g), compensating the vessels weight at the stationary point of floating. With some elevation of the ship z_g or similarly an actual wave elevation η, the hydrostatic force F_{hs} is calculated as

$$F_{hs} \approx mg + \rho_{hy}gA_W(\eta - z_g), \tag{9}$$

with gravity acceleration g, water density ρ_{hy} and the water plane area at equilibrium A_W. The integration of the waterline surface changes is neglected leading to the above linear relation.

The point where the force acts depends on the height difference of the buoyancy point above the center of gravity $h_{hs,0}$ and the second moments of water plane area $I_{L/T}$.

For the side shift, we get

$$\mathbf{r}_{hs} \approx -\left(\frac{\rho_{hy}}{m} I_L + h_{hs,0}\right) \sin(\varphi_{eff}) \mathbf{e}_y + \left(\frac{\rho_{hy}}{m} I_T + h_{hs,0}\right) \sin(\theta_{eff}) \mathbf{e}_x, \tag{10}$$

again neglecting the change of the water plane due to roll and heave. The first part of each component respects the stabilizing influence of the hull shape of the ship, the second the influence of the static deviation from the center of gravity. The second moment of water plane area I_T for the pitch movement is at a larger order than I_L for the roll movement due to the lengthiness of common boat shapes. Therefore, pitch angles stay small especially in the absence of waves.

The effective hydrostatic roll angle has to be referred to the water surface including waves. With the wave elevation field $\eta(\mathbf{x}, t)$ it becomes $\varphi_{eff} = \varphi - \arctan\left(\frac{\partial \eta}{\partial y}\right)$ and $\theta_{eff} = \theta + \arctan\left(\frac{\partial \eta}{\partial x}\right)$, where it is assumed, that the wave length is large compared to the ship length (4 m), which should be the case for offshore ocean waves with significant amplitudes.

The buoyancy force changes the direction according to the water surface, and partially acts in horizontal directions, again linearized to $\mathbf{F}_{hs} = F_{hs}\mathbf{e_z} - F_{hs}\frac{\partial \eta}{\partial x}\mathbf{e_x} - F_{hs}\frac{\partial \eta}{\partial y}\mathbf{e_y}$.

2.5 Damping

Lastly, we model linear constant damping of the boat, though the addition of a quadratic term will be more appropriate. The used damping force and moment are $\begin{pmatrix}\mathbf{F}_D & \mathbf{T}_D\end{pmatrix} = -\mathbf{d}\begin{pmatrix}\mathbf{v} & \mathbf{\Omega}\end{pmatrix}$,where we use decoupled, rough estimates for the damping parameters $\mathbf{d}$.

2.6 Inputs

For a simulation, we have to specify all external variables like wind and waves as well as control variables, rudder and sail angles. The first are part of the simulation environment, while the latter have to be chosen by a boat controller.

For the rudder, a limitation (±35°) is used. For the choice of the rudder angle, the flow separation at high angles of attack of the rudder has to be avoided or respected during controller design. For the sails, the possible angle is geometrically restricted by shrouds (±90°). A stronger constraint is imposed by the limited rate of change for which we introduce a first order model in the next part. Additionally, for practical reasons of energy management, the sail angle should not be changed too often. An adaption seems to be necessary mainly at maneuvers and changes in heading and wind. In contrast, the rudder is important to control the heading and reducing the impact of disturbances like waves or wind gusts. We choose a sample time of 100 ms, but its value may be adapted according to energy resources and control requirements.

2.6.1 Limited actuator speeds

It is not possible to set sail and rudder angles immediately, instead changing them requires time and energy. The actuators have to change the rope length (sail) or the rudder position. To respect this in a simple way, we chose a first order linear behavior determined by some time constant τ describing the typical time for the rudder respectively sail to be set. This temporal behavior results from the actuator position controllers. Additionally, the absolute change rate may be limited due to a maximum actuator speed.

The time constant of the rudder is assumed to be short (0.1 s), while for the sail we expect a larger time constant due to hardware design (e.g. 3 s). For the sails, the actuation by ropes implies another effect: Its angle can change very fast during jibing or tacking. It is rather the maximum absolute angle that is chosen by the controller setting the length of the rope. The side respectively sign of the sail angle is chosen by the relative wind.

2.7 Implementation and numeric integration of the differential equations

For the simulation, the differential equations of the dynamics model (1)-(4) have to be integrated departing from the initial state values $\mathbf{x_0}$. Therefore we use the Dormand Prince method, a standard Runge-Kutta method with adaptive step size. For a simulation time of 120 seconds, 2 seconds computation time is needed on 2 GHz single CPU computation.

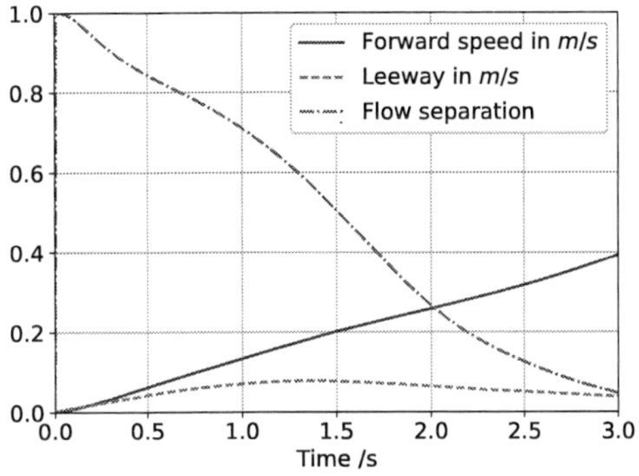

Figure 2: Boat accelerating from zero speed. Forward speed v_x and leeway v_y are both rising, until the overall speed augments and the keel compensates the side force. The high initial leeway drift angle leads to flow separation.

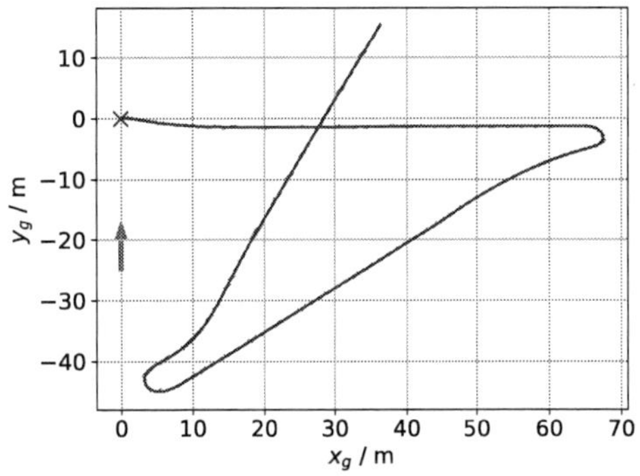

Figure 3: Simulated ship trajectory: Gaining speed tacking and jibing; boat starts at blue cross, wind direction is marked by the green arrow.

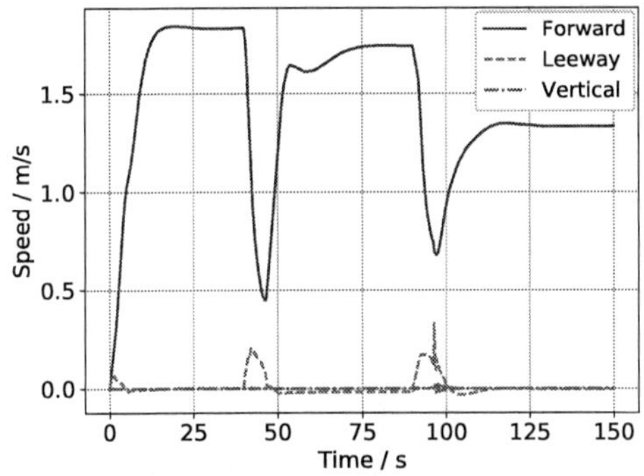

Figure 4: Speed of the boat in the different directions: Horizontal speed is almost decoupled, hence close to zero in the absence of waves.

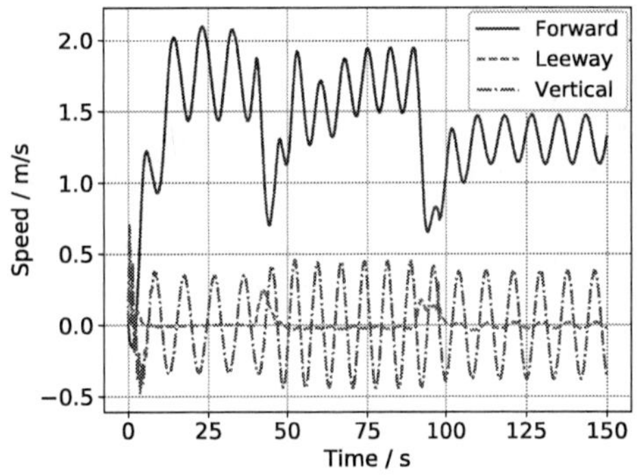

Figure 5: Same simulation scenario but with waves: Ship speeds for a wave field with 1 m wave height at a wave length of 100 m, arriving from right.

The implementation of our simulation setup is available at `https://github.com/simko96/stda-sailboat-simulator`.

3 Simulation

The simulator should be able to reproduce main effects of the sailboat dynamics to deliver useful data for testing software components like controllers or path planning. One interesting situation is the departure from zero speed, where the effects of rudder and keel differ from normal operation (here the inertia dominates). Other interesting situations include state changes during maneuver execution and the influence of bigger waves. The simulation serves as data generation to test our modules like heading controller, path planning or filter algorithms. However, so far we do not have empirical data to validate its performance. Instead we qualitatively examine for plausibility.

3.1 Scenario: gaining speed, tacking and jibing

Our scenario consists of the boat accelerating at departure from zero speed, while the wind comes from south with 5m/ s (right angle to initial heading), followed by tack and jibe maneuvers (Figure 3). Therefore, we use a feedback control for the heading through the rudder, that is derived from a simplified nonlinear dynamics model for the yaw motion, and a feed forward control of the sails.

The first part of the scenario is to gain speed (Figure 2). At departure, there is no water flow which leads to a high initial leeway (acceleration according to the sail force), and the starting flow separates. While the boat

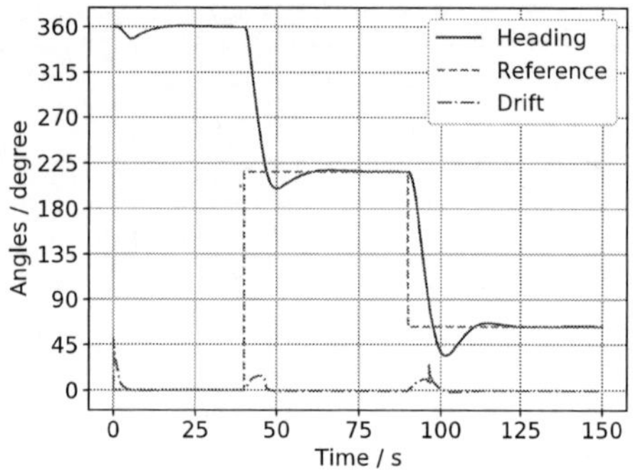

Figure 6: Plot of the controlled heading angle, its reference and the leeway drift angle compensated by the controller..

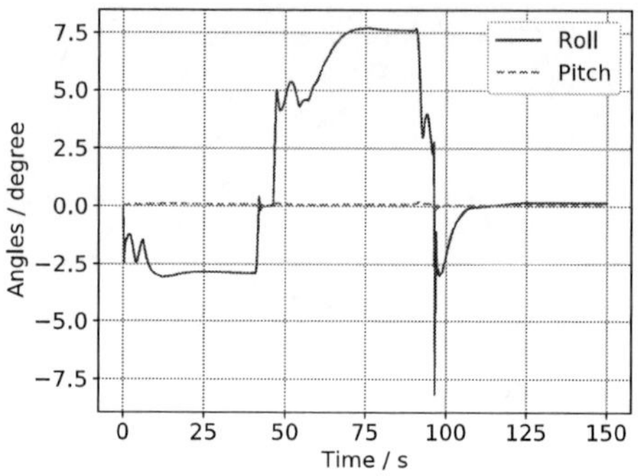

Figure 7: Corresponding angles in roll and pitch, due to the high longitudinal stability (geometric inertia), the pitch angle remains small.

speeds up, the keel force raises until it compensates for the side force of the sails.

At tacking, the speed reduces due to the missing propulsion when crossing the wind. Afterwards, the boat speeds up at the new heading. For jibing, the speed decrease is less pronounced (Figure 4). During both maneuvers, the roll angle changes the sign. The pitch angle remains negligible during the scenario (Figure 7) and becomes larger but still small when introducing waves (not shown). The heading (controlled by use of the rudder) is shown in Figure 6.

With waves of 1 m height coming from east ($-x_g$-direction), the base trajectory looks similar. Oscillations induced by the waves are superposed on all states, exemplary shown for the speeds in Figure 5.

3.2 Visualization by hardware model

Another application is the data generation for our movable hardware model of the boat shape that serves for demonstration and visualization issues (Figure 8). It is eligible when presenting our project, e.g. at fairs, and useful to test interfaces between the different components, software, electronics and actuators.

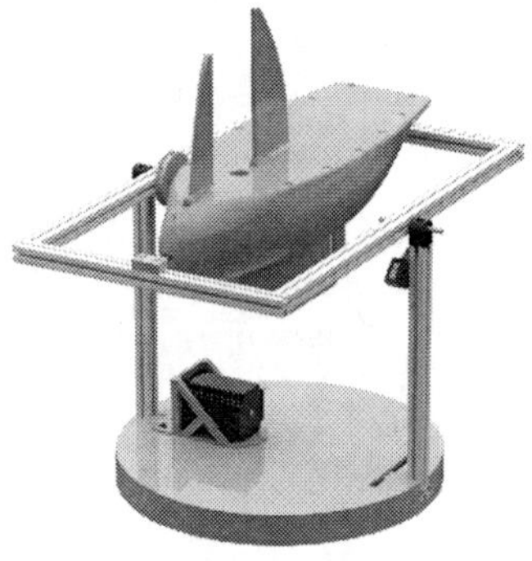

Figure 8: Demonstration model of a sailing boat.

4 Conclusion

In this paper, we derived the differential equations describing the dynamics of a sailboat, by considering the forces generated by both fluid flows, wind and the relative water stream.

This model is used, to simulate the sailboats motion, particularly for maneuvers like tacking or jibing. The simulation results seem plausible and well suitable for our demonstration model. Further, it is useful for testing our control and path planning algorithm.

However, we still need to record real data, to identify parameters like damping and compare and evaluate the simulation quality.

References

Alves, C. (2010). Sailbot: Autonomous marine robot of eolic propulsion. Master's thesis, Instituto Superior Tecnico, Universidade Tecnica de Lisboa.

Fossen, T. (2005). A nonlinear unified state-space model for ship maneuvering and control in a seaway. *Journal of Bifurcation and Chaos*, 15:2717–2746.

Korotkin, A. (2009). *Added Masses of Ship Structures.* Springer.

Masuyama, Y. and Fukasawa, T. (2011). Tacking simulation of sailing yachts with new model of aerodynamic force variation during tacking maneuver. *Journal of Sailboat Technology*, 119.

Philpott, A. B., Sullivan, R. M., and Jackson, P. S. (1993). Yacht velocity prediction using mathematical programming. *European Journal of Operational Research*, 67:13 – 24.

Roncin, K. and Kobus, J. (2004). Dynamic simulation of two sailing boats in match racing. *Sports Engineering*, 7:139–152.

Saoud, H., Hua, M., Plumet, F., and Amar, F. (2013). Modeling and control design of a robotic sailboat. *Robotic Sailing*.

Spurk, J. and Aksel, N. (2010). *Stroemungslehre*. Springer.

Backstepping Control of an Autonomous Catamaran Sailboat

Helmi Abrougui
U.R Automatic Control & Marine Robotics
Naval Academy, Tunisia
helmiabrougui@yahoo.fr

Samir Nejim
U.R Automatic Control & Marine Robotics
Naval Academy, Tunisia
samir.nejim@centraliens.net

Abstract

Automatic control of sailing boats has continuously evolved since it allows the increase of sailing safety and the improvement of cruise speed. Hence, the main problem with sailing resides in the large heel angle and important roll motion that can be caused by strong wind. In order to overtake these risks and to ameliorate the sailboat manoeuvrability and speed, a solution is to design sailboats with two or three hulls; referred, respectively, as catamaran and trimaran. This paper presents an autopilot design for an autonomous catamaran sailing vessel. A four degree of freedom dynamic model of the vehicle is firstly described using Newton's second law and kinematic equations. Due to high nonlinearity of the mathematical dynamic model of the catamaran sailboat, a nonlinear heading controller based on backstepping method is developed to stabilize the boat heading while tracking waypoints. Finally, simulation results are carried out to show the effectiveness of the proposed approach and the behaviour of the overall system.

1 Introduction

Sailing is an efficient navigation technic that uses wind kinetic energy but little to no electric energy to navigate. Therefore, sailboats are well suited for long operations such as monitoring of maritime area, oceanographic research and Microtransat challenge (Brière, 2006). In this context, many autonomous sailboat projects have been launched throughout the world over the last decade such as the AVALON, the AROO and the FASt sailboats project (Erckens et al, 2010) (Neal, 2006) (Alves et al, 2008).
The sailboat propulsion depends on the wind speed and direction. The sail is used to create forward propulsion for the sailing vessel. Depending on the polar diagram of the sailboat, there is an optimal sail angle that gives the highest forward linear speed but it can cause an important heel angle that decreases the boat stability and safety. Hence, we need an effective control of the sail and the rudder angle to reduce the roll motion. Authors (Wille et al, 2016) developed a Linear Quadratic Regulator (LQR) for controlling the momentum created by the sail. This controller reduces the roll motion so, it increases robustness and safety. Another solution for reducing roll motion is to design catamaran sailboats. Due to its two hull the catamaran sailboat is more stable than the monohull sailboat.

Figure 1: 3D printing catamaran sailboat

In: S. M. Schillai, N. Townsend (eds.): Proceedings of the International Robotic Sailing Conference 2018, Southampton, United Kingdom, 31-08-2018.

Classical techniques of Lagrangian and Newtonian mechanics are the most used methods for determining the mathematical model of a catamaran sailing boat. An example of a dynamic model with six Degree of Freedom 6-DoF is presented in (Furrer, 2010) using Fossen approach (Fossen, 2002).

Several control techniques have been applied in sailing vessel. The most used regulators are the Proportional Integral Derivative PID controllers (Cruz et al, 2014) and (Ramirez, 2012). This control law is designed using Nomoto's first order model. In addition, a Mamdani type fuzzy inference systems was used by authors (Stelzer et al, 2007) to control both actuators for sail and rudder. Another fuzzy-based controller has been developed in (Gomes et al, 2015) and (Abril et al, 1997) where the rudder angle was calculated via the heading and the desired angular velocity. Xiao and Jouffroy (Xiao et al, 2014) designed a nonlinear heading controller for a 4-DoF monohull sailboat model using backstepping technics. The basic idea behind this method is to break down the design problem of the full system into a sequence of sub-problems on lower order systems, and recursively use some states as "virtual input" to obtain the intermediate control laws with the Control Lyapunov Function (CLF). The main advantage of backstepping control is the guarantee of system stability.

In this paper, due to the high nonlinearities of the developed dynamic model of the catamaran sailboat, an autopilot was designed based on backstepping approach in order to steer the vehicle towards a specific target. For a given wind direction, the sail position is defined as a direct function of the boat heading.

In the following section, a dynamic model of a sailing catamaran with four degree of freedom is described. In section III, the nonlinear heading controller was designed and tested with the developed dynamic model. Finally, some simulation results were carried out to illustrate and to evaluate the studied approach.

2 System Dynamics

The dynamic model of the catamaran sailboat presented in this paper is inspired from previous researches conducted in (Jaulin et al, 2013) and (Wille et al, 2016). One of the main contributions of this paper consists of the improvement of the dynamic model by taking into consideration the roll and the sway motion. Therefore, the resulting dynamic model has 4-DoF instead of three. This dynamic model of the sailing catamaran is derived under the following assumptions:

- The sail and the rudder are modeled as rigid foils.
- Added mass coefficients (Fossen, 2002) are modeled as constants.
- The sailboat is assumed to evolve in calm waters.

The modeled catamaran sailing boat is presented in Figure 2 and Appendix B, where the North-East-Down NED coordinate system (X, Y, Z) is treated as inertial reference frame $(n - frame)$ and the (x_b, y_b, z_b) is the body fixed frame $(b - frame)$ connected to the boat.

The latter is the reference frame attached to the sailboat. It rotates with angular velocity $W = (p\ q\ r)^T$ relative to the $(n - frame)$. Its origin is assumed to coincide with the sailboat's center of gravity G.

The sailboat linear velocity in $(b - frame)$ is $V = (u\ v\ w)^T$.

The sailboat is assumed to be rigid and 4-DoF are considered, after excluding both heaving and pitching motions $q = w = 0$. (See Figure 1)

$\boldsymbol{v} = (u\ v\ r\ p)^T$ is the velocity vector in the $(b - frame)$ and $\eta = (x\ y\ \psi\ \phi)^T$ is a vector describing, the position of the sailboat, its yaw and roll in the $(n - frame)$.

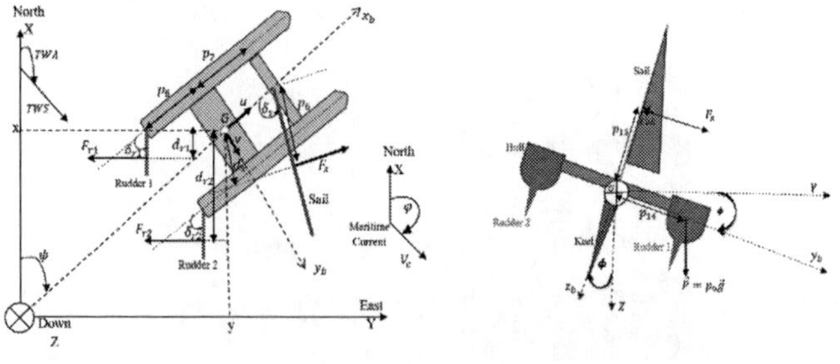

Figure 2: Top and rear view of the modelled catamaran sailboat

Vector η is then derived through a coordinate transformation (Fossen, 2002), giving the following kinematic equations of the sailboat:

$$\dot{x} = u\cos\psi - v\sin\psi\cos\phi + V_C\cos\varphi \tag{1}$$

$$\dot{y} = u\sin\psi + v\cos\psi\cos\phi + V_C\sin\varphi \tag{2}$$

$$\dot{\psi} = r\cos\phi \tag{3}$$

$$\dot{\phi} = p \tag{4}$$

Where φ and V_c are respectively the maritime current direction and speed.
The sail of the boat is inflated by the apparent wind force and consequently the sailboat advances.
According to (Xiao et al, 2014), the apparent wind AW is the vector sum of the true wind TW in $(b - frame)$ $TW^{\,b-frame}$ and the sailboat velocity $\boldsymbol{v}$. (Appendix A).
The true wind vector expressed in the $(n - frame)$ $TW^{\,n-frame}$ is given by:

$$TW^{\,n-frame} = \begin{pmatrix} TWS\cos TWA \\ TWS\sin TWA \\ 0 \end{pmatrix} \tag{5}$$

So,

$$TW^{\,b-frame} = R_2 R_1 TW^{\,n-frame} \tag{6}$$

With

$$R_1 = \begin{pmatrix} \cos\psi & \sin\psi & 0 \\ -\sin\psi & \cos\psi & 0 \\ 0 & 0 & 1 \end{pmatrix}$$

$$R_2 = \begin{pmatrix} 1 & 0 & 0 \\ 0 & \cos\phi & \sin\phi \\ 0 & -\sin\phi & \cos\phi \end{pmatrix}$$

Therefore, we get

$$\begin{aligned} AW^{\,b-frame} &= TW^{\,b-frame} - V - W \times (x_s\; y_s\; z_s)^T \\ &= \begin{pmatrix} TWS\cos(TWA - \psi) - u - r\,y_s \\ TWS\sin(TWA - \psi)\cos\phi - v - r\,x_s + pz_s \\ 0 \end{pmatrix} \\ &= \begin{pmatrix} AW^{x_b} \\ AW^{\,y_b} \\ 0 \end{pmatrix} \end{aligned} \tag{7}$$

Then

$$AWA = atan2(AW^{y_b}, AW^{x_b}) \tag{8}$$

$$\begin{aligned} AWS &= \left\| AW^{\,b-frame} \right\| \\ &= \sqrt{(AW^{x_b})^2 + (AW^{y_b})^2} \end{aligned} \tag{9}$$

The angle of attack on the sail by the direction of the apparent wind AW is equal to $(\delta_s - AWA)$ (Melin, 2015). Therefore, the aerodynamic force f_s applied on the Center of Effort CoE of the sail is equal to:

$$f_s = p_4 AWS\sin(\delta_s - AWA) \tag{10}$$

The vectorial representation of this aerodynamic force in the fixed body frame $(b - frame)$ is:

$$F_s = \begin{pmatrix} f_s^{x_b} \\ f_s^{y_b} \end{pmatrix} = \begin{pmatrix} f_s\sin\delta_s \\ f_s\cos\delta_s \end{pmatrix} \tag{11}$$

On the other hand, the angle of attack on the rudders by the apparent water velocity is equal to δ_r (Melin, 2015). By assuming that the apparent water and the sailboat speeds are equal, the water generates a hydrodynamic force on each rudder which is equal to:

$$\begin{cases} f_{r1} = p_5 u^2 \sin\delta_{r1} \\ f_{r2} = p_5 u^2 \sin\delta_{r2} \end{cases} \tag{12}$$

With

$$\delta_{r1} = \delta_{r2} = \delta_r \tag{13}$$

Thus, in what follows the hydrodynamic force applied on the sailboat is supposed created by one rudder situated in $(x_b - axis)$ and equal to:

$$f_r = f_{r1} + f_{r2} = 2p_5u^2 \sin\delta_r \tag{14}$$

The vectorial representation of this hydrodynamic force in the fixed body frame $(b - frame)$ is then

$$F_r = \begin{pmatrix} f_r^{x_b} \\ f_r^{y_b} \end{pmatrix} = \begin{pmatrix} -f_r \sin\delta_r \\ -f_r \cos\delta_r \end{pmatrix} \tag{15}$$

For simplicity reasons, the sailboat is affected by some tangential friction forces $-p_1u^2$ and $-p_2v$ respectively applied on $(x_b - axis)$ and $(y_b - axis)$ also it is affected by an angular friction force $-p_3r$ applied around $(z_b - axis)$.

In addition, the applied forces on the sailboat are F_s and F_r. Therefore, according to Newton's second law of motion applied in the $(b - frame)$, we have:

$$\begin{aligned} (p_9 - X_{\dot u})\dot u &= f_s^{x_b} + f_r^{x_b} - p_1u^2 && (16)\\ &= f_s \sin\delta_s - f_r \sin\delta_r - p_1u^2 \\ (p_9 - Y_{\dot v})\dot v &= f_s^{y_b} + f_r^{y_b} - p_2v && (17)\\ &= f_s \cos\delta_s - f_r \cos\delta_r - p_2v && (18)\\ (p_{10} - N_{\dot r})\dot r &= d_sf_s - d_rf_r - p_3r \end{aligned}$$

With

$$d_s = p_6 - p_7 \cos\delta_s$$
$$d_r = p_8 \cos\delta_r$$

By adding the Coriolis effect caused by both rigid body and added mass (Wille et al, 2016) we get:

$$(p_9 - X_{\dot u})\dot u = f_s \sin\delta_s - f_r \sin\delta_r + vr(p_9 - Y_{\dot v}) - p_1u^2 \tag{19}$$
$$(p_9 - Y_{\dot v})\dot v = f_s \cos\delta_s - f_r \cos\delta_r + ur(X_{\dot u} - p_9) - p_2v \tag{20}$$
$$(p_{10} - N_{\dot r})\dot r = (p_6 - p_7 \cos\delta_s)f_s - p_8 \cos\delta_r f_r + uv(Y_{\dot v} - X_{\dot u}) - p_3r \tag{21}$$

According to Le Bars (Le Bars et al, 2013) the roll motion is supposed to be pendulum:

$$\dot p = \frac{z_sf_s \cos\delta_s \cos\phi - p_{13}p_9g \sin\phi - p_{12}p}{p_{11} - K_{\dot p}} \tag{22}$$

Therefore, the 4-DoF state equations, which describe the dynamics of the sailboat, are:

$$\begin{cases} \dot x = u\cos\psi - v\sin\psi\cos\phi + V_C\cos\varphi \\ \dot y = u\sin\psi + v\cos\psi\cos\phi + V_C\sin\varphi \\ \dot\psi = r\cos\phi \\ \dot\phi = p \\ \dot u = \dfrac{(f_s \sin\delta_s - f_r \sin\delta_r + vr(p_9 - Y_{\dot v}) - p_1u^2)}{(p_9 - X_{\dot u})} \\ \dot v = \dfrac{(f_s \cos\delta_s - f_r \cos\delta_r + ur(X_{\dot u} - p_9) - p_2v)}{(p_9 - Y_{\dot v})} \\ \dot r = \dfrac{((p_6 - p_7\cos\delta_s)f_s - p_8 f_r \cos\delta_r + uv(Y_{\dot v} - X_{\dot u}) - p_3r)}{(p_{10} - N_{\dot r})} \\ \dot p = \dfrac{z_sf_s \cos\delta_s \cos\phi - p_{13}p_9g \sin\phi - p_{12}p}{p_{11} - K_{\dot p}} \end{cases} \tag{23}$$

This system state space (23) is highly nonlinear and it has the following compact notation; $\dot X = f(X, U, B)$
With:

$X = (x\; y\; \psi\; \phi\; u\; v\; r\; p)^T$ the state vector.

$U = \begin{pmatrix} \delta_s \\ \delta_r \end{pmatrix}$ the input vector.

B, disturbance due to maritime current.

The two actuators, sail and rudder, have physical limitations. In this dynamic model, these limitations are expressed by setting the maximum sail opening angle to $\delta_s^{max} = 90°$ and the maximum angular velocity to 30 fl/s. In (Santos et al, 2016) and (Xiao et al, 2014), the control of the sail opening angle was assured by setting

the length of the rope which is attached to the boom. However, in this paper, it is assumed that the sail is rotated by a servomotor built in the mast. Therefore, the sail and servo angle were assumed to be equal.
The rudder will follow similar limitations, $\delta_r^{max} = 45°$ for maximum rudder angle and 30 fl/s for its maximum angular velocity.

3 Autopilot Design

To make an autonomous catamaran sailboat follow a given path or reach a target position(x_d, y_d), a controller with two levels is designed as presented in Figure 3. The first one is a high-level controller. It generates the desired heading ψ^{ref} and the sail opening angle δ_s for a sailing trip. The second controller is the autopilot. It usually sails the boat to the desired heading. In this paper, we will focus on the autopilot design. Regarding the navigation strategy (high-level controller), we will use the waypoint tracking proposed in (Jaulin, 2014) with minor improvements.

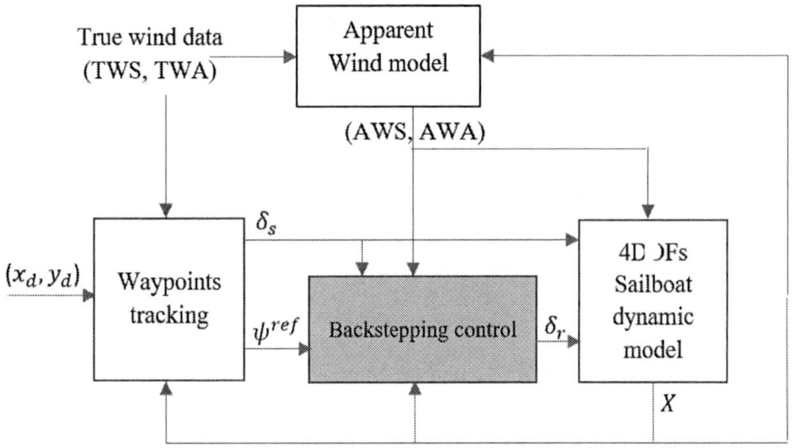

Figure 3: Proposed control scheme

3.1 Heading Control

To use the backstepping techniques we consider again the subsystem described by the two differential equations (3) and (21).
The heading error is: $e_1 = \psi^{ref} - \psi$, its dynamics is written as $\dot{e}_1 = -r\cos\phi$ since $\dot{\psi}^{ref} = 0$
Let the Control Lyapunov Function be defined as

$$V_1 = \frac{1}{2}e_1^2 \tag{24}$$

We first choose r as the virtual control input to converge the system toward $e_1 = 0$ with a stabilizing function $g_1(e_1, \phi)$ such that $\dot{V}_1 < 0$ for all $e_1 \neq 0$.
We have

$$\dot{V}_1 = e_1\dot{e}_1 = -re_1\cos\phi \tag{25}$$

By taking $r = g_1(e_1, \phi) = \frac{k_1 e_1}{\cos\phi}$ we obtain $\dot{V}_1 = -k_1 e_1^2 < 0$
Where k_1 is a positive design constant.
However, since r is not the system input, we can achieve $r = g_1(e_1, \phi) = \frac{k_1 e_1}{\cos\phi}$ only with an error that is represented by the new variable error

$$e_2 = g_1(e_1, \phi) - r \tag{26}$$

The new dynamic errors are:

$$\begin{cases} \dot{e}_1 = -r\cos\phi = \cos\phi\left(e_2 - \dfrac{k_1 e_1}{\cos\phi}\right) = e_2\cos\phi - k_1 e_1 \\ \dot{e}_2 = \dfrac{k_1}{\cos^2\phi}(e_1 p\sin\phi - r\cos^2\phi) - h(\psi, r, u, v, \delta_r, f_s) \end{cases} \quad (27)$$

with

$$h(\psi, r, u, v, \delta_r, f_s) = \dot{r} \quad (28)$$

So, the new control Lyapunov function V_2 can be written as

$$V_2 = \frac{1}{2}e_1^2 + \frac{1}{2}e_2^2 \quad (29)$$

Then

$$\dot{V}_2 = e_1\dot{e}_1 + e_2\dot{e}_2 \quad (30)$$

given that

$$e_1\dot{e}_1 = e_1(e_2\cos\phi - k_1 e_1) \quad (31)$$

$$e_2\dot{e}_2 = e_2\left(\frac{k_1}{\cos^2\phi}(-r\cos^2\phi + e_1 p\sin\phi) - h(\psi, r, u, v, \delta_r, f_s)\right) \quad (32)$$

$\dot{V}_2$ can be rewritten as:

$$\dot{V}_2 = e_1 e_2\cos\phi - k_1 e_1^2 - e_2 h(\psi, r, u, v, \delta_r, f_s) - e_2 k_1 r + \frac{e_2 e_1 k_1 p}{\cos^2\phi}\sin\phi \quad (33)$$

By choosing

$$h(\psi, r, u, v, \delta_r, f_s) = e_1\left(\cos\phi + \frac{k_1 k_2}{\cos\phi} + \frac{k_1 p\sin\phi}{\cos^2\phi}\right) - r(k_1 + k_2) \quad (34)$$

We get

$$\dot{V}_2 = -k_1 e_1^2 - k_2 e_2^2 < 0 \quad \forall \quad e_1 \neq 0,\ e_2 \neq 0 \quad (35)$$

With $k_2 \in \Re^{*+}$

Using equations (14), (28) and (34) we obtain:

$$\delta_r = 0.5\,sin^{-1}\left(\frac{(p_6 - p_7\cos(\delta_s))f_s + uv(Y_{\dot{v}} - X_{\dot{u}}) - r\left(p_3 - (p_{10} - N_{\dot{r}})(k_1 + k_2)\right) - e_1(p_{10} - N_{\dot{r}})\left(\cos\phi + \frac{k_1 k_2}{\cos\phi} + k_1 p\frac{\sin\phi}{\cos^2\phi}\right)}{0.5 p_5 p_8 u^2}\right) \quad (36)$$

With

$$\begin{cases} \delta_r < |\delta_r^{max}| \\ u \neq 0 \ \ \forall\ X \in \Re^8 \end{cases}$$

3.2 Navigation Strategy

Based on the polar diagram[1] given in Figure 5 and using the target position (x_d, y_d), the system state space and the wind direction, the high-level controller which is inspired from (Jaulin, 2014) generates the desired heading ψ^{ref} and the sail opening angle δ_s.

As we know, a sailing boat cannot sail directly against the wind direction because in this case the sail will be luffing in the breeze and cannot generate any propulsive power.

The polar diagram (see Figure 5) shows the sector of directions that cannot be navigable by sailboats. It is called a no-go-zone. The size of the no-go zone, referred to as the no-go-heading, will differ based on characteristics of the particular sailboat.

To avoid this no-go-zone, the boat has to sail in Zig-Zag course and execute tack[2] manoeuver so, the high-level controller must generate desired heading $\psi^{ref} \notin]\mathrm{TWA} + \pi - \xi, \mathrm{TWA} - \pi + \xi[$.

[1] is the set of all pairs (ψ, u) that can be reached by the sailboat when it navigates.

[2] The sailboat tacks, that is sails on alternating sides of the wind and therefore advances towards the wind. This is the most complex case, testing the interaction between all parts of the model, especially rudder and sail forces.

So, in the case where the sailboat has to navigate with maximum speed on an upwind navigation ($TWA = \pi + \psi$) the following set of headings are used.

$$\psi_1^{ref} = TWA + \pi - \xi - \beta \quad \text{and} \quad \psi_2^{ref} = TWA - \pi + \xi + \beta \tag{37}$$

For downwind sailing, the desired heading given by the high-level controller is defined by the following $atan2$ function.[3]:

$$\psi_3^{ref} = atan2(y - y_d, x - x_d) \tag{38}$$

which allows the sailboat to navigate directly toward its destination.

Using the algorithm given in (Santos, 2016) and presented on Figure 4, the overall system was simulated to evaluate the control law developed in this work.

In what follows, the wind is simulated coming from the North ($TWA = -\frac{\pi}{2}$) and its speed is equal to 10 m/s. Therefore the sail opening angle is given by the following equation, it depends on the sailboat heading ψ:

$$\delta_s = \pi\left[\frac{\psi}{2\pi} + \frac{1}{4}\right] + \frac{\pi}{4} - \frac{\psi}{2} \tag{39}$$

Using equations (27) and (34) the error dynamics is

$$\begin{cases} \dot{e}_1 = e_2 \cos\phi - k_1 e_1 \\ \dot{e}_2 = -e_1\left(\cos\phi + \dfrac{k_1 k_2}{\cos\phi}\right) + r k_2 \end{cases} \tag{40}$$

Hence, the controller (36) is tuned with $k_1 = k_2 = 1$ using the pole placement method applied to equation system (40).

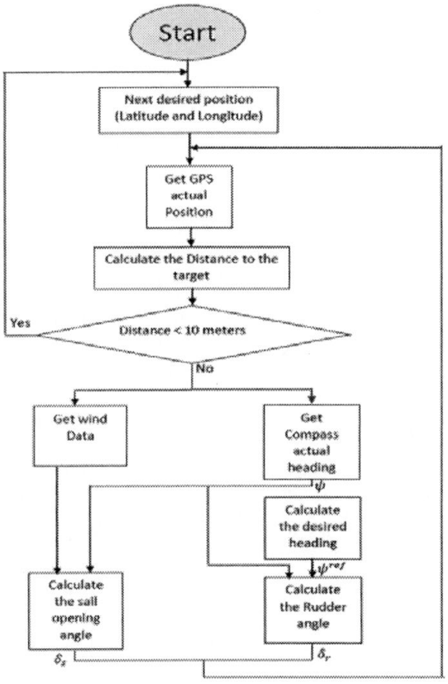

Figure 4: Algorithm used to control the sailboat (Guidance, Navigation and Control)

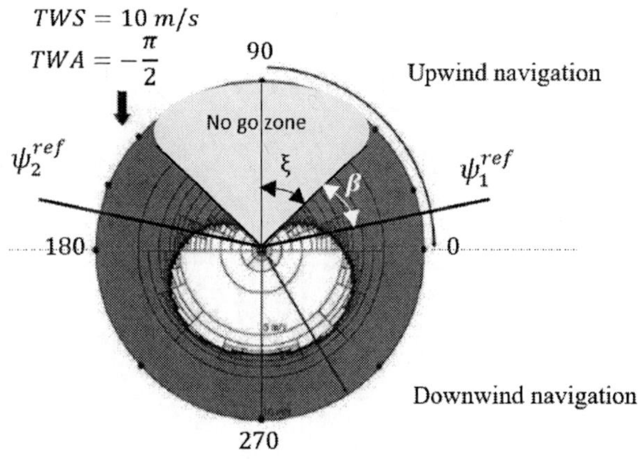

Figure 5: Sailboat polar diagram $\xi = \frac{\pi}{4}$, $\beta = \frac{\pi}{12}$

[3] $atan2(y, x) \in [-\pi, \pi]$ is the four-quadrant inverse tangent.

4 Simulation and Results

Four set of simulations have been studied in order to show the path followed by the sailboat against the wind direction. In these simulations, the target is situated in position $(x_d = 0\,;y_d = 0)$. The simulations started with initial values $X(t=0) = (x_i\ y_i\ \psi_i\ 0\ 8\ 0\ 0\ 0\ )^T$ with $i \in \{1,2,3,4\}$ and stopped when the target position was reached with a distance equal to ten meters.
The proposed control law allows the sailboat to reach the target position. Figure 6 clearly shows that the boat sails on Zig-Zag course when the target is situated against the true wind direction $i \in \{3,4\}$. However, it goes directly to the target position in downwind case $i \in \{1,2\}$. We can conclude that there is a good synchronization between the sail opening angle Figure 8 and the rudder control Figure 7 in order to execute the upwind tack maneuver. Figure 9 shows that the heading $\psi(t)$ eventually converges to the desired heading $\psi^{ref}(t)$ with an error $\psi^{ref}(t) - \psi(t) = 0$.

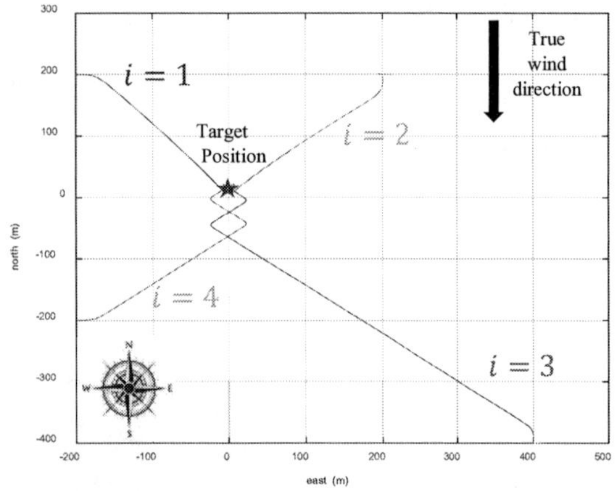

Figure 6: Paths followed by the catamaran sailboat
$(x_d = 0\,, y_d = 0)$

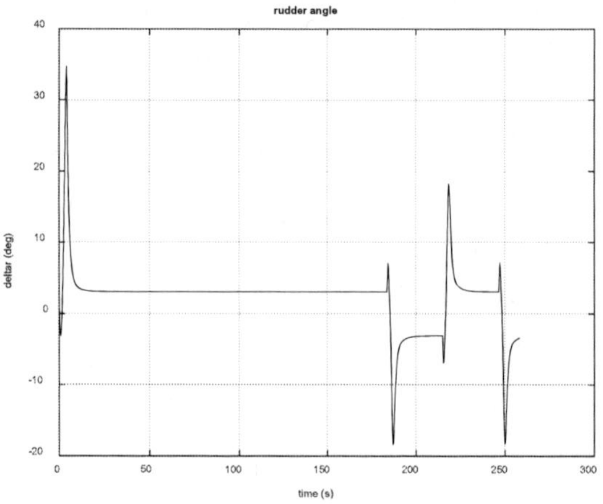

Figure 7: Time evolution of the rudder angle control $(i = 3)$

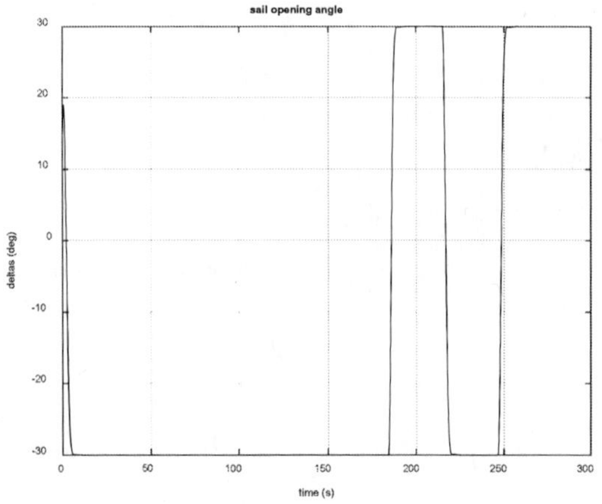

Figure 8: Time evolution of the sail opening angle $(i = 3)$

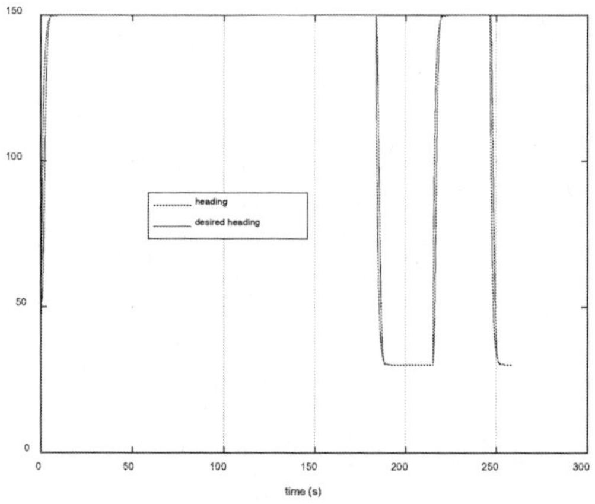

Figure 9: Time evolution of the desired heading ψ^{ref} and the sailboat heading ψ $(i = 3)$

5 Conclusion

In this paper, a 4-DoF mathematical model describing the dynamic motion of a catamaran sailboat was presented. This dynamic model is considered a satisfactory model to represent the vehicle dynamic motion in the horizontal plane. The backstepping control method is used to develop a nonlinear heading control for the sailing boat. This controller works as intended. It allows to keep the sailboat on a predefined heading while waypoint tracking. The control system stability was proved using Control Lyapunov Function. As a future work in this area, it is suggested to identify the considered 3D printed catamaran sailboat (Figure 1) model parameters using the temporal variation data of the yaw angle (Casado et al, 2005). Afterwards, the implementation of the developed control law in real time will be done.

References

Cruz, N. A., & Alves, J. C. (2014). Navigation performance of an autonomous sailing robot. *In Oceans-St. John's*, (pp. 1-7). *IEEE*.

Erckens, H., Büsser, G. A., Pradalier, C., & Siegwart, R. Y. (2010). Navigation strategy and trajectory following controller for an autonomous sailing vessel. *IEEE RAM*, 17, 47-54.

Jaulin, L., & Le Bars, F. (2013). A simple controller for line following of sailboats. *In Robotic Sailing* (pp. 117-129). *Springer, Berlin, Heidelberg*.

Jaulin, L. (2004). Modelisation et commande d'un bateau à voile. In *Proceedings of 3rd Conference Internationale Francophone d'Automatique, Douz, Tunisie*.

Furrer, F. (2010). Developing a simulation model of a catamaran using the concept of hydrofoils. *ETH, Zurich*.

Romero-Ramirez, M. A. (2012). Contribution à la commande de voiliers robotisés (*Doctoral dissertation, Paris 6*).

Stelzer, R., Proll, T., & John, R. I. (2007). Fuzzy logic control system for autonomous sailboats. *In Fuzzy Systems Conference*, 2007. *FUZZ-IEEE. IEEE International* (pp. 1-6).

Xiao, L., & Jouffroy, J. (2014). Modeling and nonlinear heading control of sailing yachts. *IEEE Journal of Oceanic engineering*, 39(2), 256-268.

Gomes, L., Santos, M., Pereira, T., & Costa, A. (2015). Model-based development of an autonomous sailing Yacht controller. *In Autonomous Robot Systems and Competitions* (ICARSC), *IEEE International Conference on* (pp. 103-108). IEEE

Wille, K. L., Hassani, V., & Sprenger, F. (2016). Roll stabilization control of sailboats. *IFAC-Papers OnLine*, 49(23), 552-556.

Santos, D., Silva Junior, A. G., Negreiros, A., Vilas Boas, J., Alvarez, J., Araujo, A., & Gonçalves, L. M. (2016). Design and implementation of a control system for a sailboat robot. *Robotics*, 5(1), 5.

Casado, M. H., & Ferreiro, R. (2005). Identification of the nonlinear ship model parameters based on the turning test trial and the backstepping procedure. *Ocean Engineering*, 32(11-12), 1350-1369.

Wille, K. L., Hassani, V., & Sprenger, F. (2016). Modeling and Course Control of Sailboats. *IFAC-PapersOnLine*, 49(23), 532-539.

Le Bars, F., Jaulin, L., & Ménage, O. (2013). Suivi de ligne pour un voilier: application au robot voilier autonome VAIMOS pour l'océanographie. *In Journées des Démonstrateurs* (p. xx).

Fossen, T. I. (2002). Marine Control System-Guidance, Navigation and Control of Ships, Rigs and Underwater Vehicles. *Marine Cybemetics*.

Neal, M. (2006). A hardware proof of concept of a sailing robot for ocean observation. *IEEE Journal of Oceanic Engineering*, 31(2), 462-469.

Alves, J. C., & Cruz, N. A. (2008). Fast-an autonomous sailing platform for oceanographic missions. *In OCEANS* (pp. 1-7). *IEEE.*

Melin, J. (2015). Modeling, control and state-estimation for an autonomous sailboat.

Brière, Y. (2006). The first microtransat challenge. *Google Scholar.*

Abril, J., Salom, J., & Calvo, O. (1997). Fuzzy control of a sailboat. *International Journal of Approximate Reasoning*, 16(3-4), 359-375.

Appendix A. Wind velocity representation

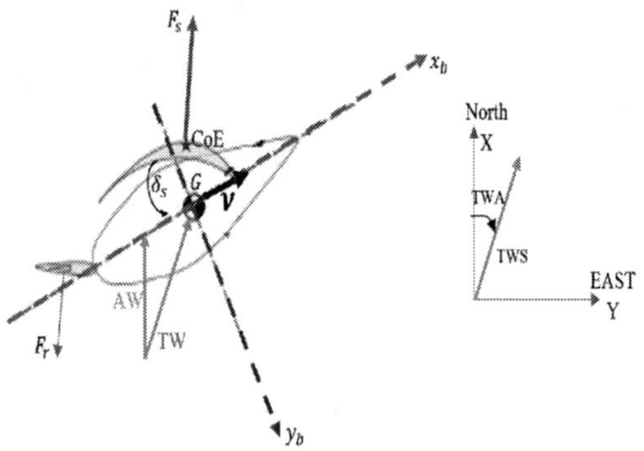

Appendix B. Variable description

Notation	Description
TW, AW	-true and apparent wind vector.
TWS, TWA	-true wind speed and direction.
AWS, AWA	-apparent wind speed and direction
V_C, φ	-maritime current speed and direction
(x, y)	-north east position
(x_s, y_s, z_s)	-coordinate of the CoE of the sailboat in (b-frame).
ψ, ϕ	-yaw and roll angle in the (n-frame).
δ_s	-sail opening angle in the (b-frame).
δ_{r1}, δ_{r2}	-rudder angle in the (b-frame).
u, v	-surge and sway motion in the (b-frame).
r, p	-angular velocity in yaw and roll (b-frame).
f_s	-aerodynamic force of the wind applied on the sail
f_{r1}, f_{r2}	-hydrodynamic forces of the water applied on the left and right rudder in the (b-frame).
g	-gravity.
p_1, p_2	-water friction in (x_b_axis) and (y_b_axis).
p_3	-water angular friction.
p_4, p_5	-lift coefficient of the sail/rudder.
p_6	-distance between the mast and CoE.
p_7	-distance between the boat's center of gravity and the mast.
p_8	-distance between G and the rudder.
p_9	-total mass of the boat.
p_{10}, p_{11}	-moment of inertia Z-axis / X-axis.
p_{12}	-roll friction coefficient.
p_{13}	-length of the equivalent pendulum in roll motion (in m).
ξ	-width of the no-go-zone, this parameter depends on the sailboat characteristics.
β	-an angle which gives maximum speed when the sailboat is on an upwind navigation.
$X_{\dot{u}}, Y_{\dot{v}}, N_{\dot{r}}, K_{\dot{p}}$	-added mass in (b-frame)

Part II

Safe Navigation

Securing Navigation of Unmanned Maritime Systems

Tope Omitola
Cyber-Physical Systems Research Group
Electronics and Computer Science, University of Southampton
t.omitola@ecs.soton.ac.uk
Jon Downes
Fluid Structure Interactions
School of Engineering, University of Southampton
jon.downes@soton.ac.uk
Gary Wills
Cyber-Physical Systems Research Group
Electronics and Computer Science, University of Southampton
gbw@ecs.soton.ac.uk
Mark Zwolinski
Sustainable Electronic Technologies Group
Electronics and Computer Science, University of Southampton
mz@ecs.soton.ac.uk
Michael Butler
Cyber-Physical Systems Research Group
Electronics and Computer Science, University of Southampton
mjb@ecs.soton.ac.uk

Abstract

Unmanned Maritime Systems (UMS), such as Unmanned Surface Vehicles (USVs), are increasingly playing a critical role in expanding the undersea superiority of a nation, addressing growing challenges, such as, inter alia, in piracy, natural resource disputes, drug trafficking, weapons proliferation, as well as being highly used for science and survey missions. Autonomous capabilities in USVs can reduce the costs of reaching into distant environments and using that reach to meet a particular mission's objectives. However, to take on increased autonomy in unmanned systems, USVs will increasingly require the ability to be untethered from human interaction, and a key enabler to effecting this is accurate navigation. USVs have traditionally depended on Global Navigation Satellite Systems (GNSS), which are known to have security and safety vulnerabilites. Using systems-theoretic process analysis (STPA), this paper provides systematic analyses of the attack surfaces and of the impact of cyber attacks against the navigational aspects of Unmanned Surface Vehicles. As part of these analyses, we identify

In: S. M. Schillai, N. Townsend (eds.): Proceedings of the International Robotic Sailing Conference 2018, Southampton, United Kingdom, 31-08-2018

potential threats, vulnerabilities and attacks in the Positioning, Navigation and Timing (PNT) functionalities of USVs. These analyses can be used to drive a USV's architecture, leading to the design of more effective and secure USV operations.

1 Introduction

Over 90% of information, people, goods and services flow across the worlds oceans (Navy,). Protecting a country's residents and economic prosperity is, therefore, essential and dependent on the ability to persistently monitor ocean surface and sub-surface activities, in order to identify, classify and mitigate emerging threats. Unmanned Maritime Systems (UMS), such as Unmanned Surface Vehicles (USVs), are increasingly playing a critical role in expanding the surface and underwater superiority of a nation, and addressing growing challenges, such as, inter alia, in piracy, natural resource disputes, drug trafficking and weapons proliferation. USVs are also employed in other areas of economic life, such as (a) Maritime search and rescue, (b) Hydrologic surveys, (c) Port surveillance, (d) Underwater Inspection, (e) Naval Defence, and their greater use could save the global marine industry up to £80 billion per annum by "potential reductions in capital costs, manning costs and fuel costs" (Rolls-Royce,).

Figure 1: ASV C-Worker USV from asvglobal.com

Because of their position at the air-sea interface, USVs have the ability to relay radio frequency transmissions in air and acoustic transmissions undersea. Thus they are a key piece in the vision of networked maritime space, both in defence and civil. Figure 1 shows an example commercial USV, while figure 2 shows an example of how BP is making use of UMS in its networked maritime space (the picture shows a number of USVs, e.g. C-Worker, WaveGlider and Autonaut, plus some underwater vehicles, such as the Seaglider, working together to monitor the ocean surface and the seabed).

Autonomous capabilities in USVs can reduce the costs and risks of reaching into distant environments while using that reach to meet missions' objectives. Marine vehicles are taking on higher levels of autonomy to perform unmanned missions, therefore securing their autonomous navigation and control modules becomes increasingly important.

2 Autonomy, Navigation and Control

Autonomy incorporates "systems which have a set of intelligence-based capabilities that allow it to respond to situations that were not programmed or anticipated in the design (i.e., decision-based responses). Autonomous systems have a degree of self-government and self-directed behavior (with the humans proxy for decisions)" (Air Force Research Laboratory, 2013). USVs must be capable of avoiding ships, docks, floating debris, and navigation aids must ensure that these USVs are able to avoid these obstacles and other marine assets, and remain in navigable waters. In addition, USVs must operate in accordance with collision regulations (COLREGS)[1].

[1]http://www.collisionregs.com/MSN1781.pdf

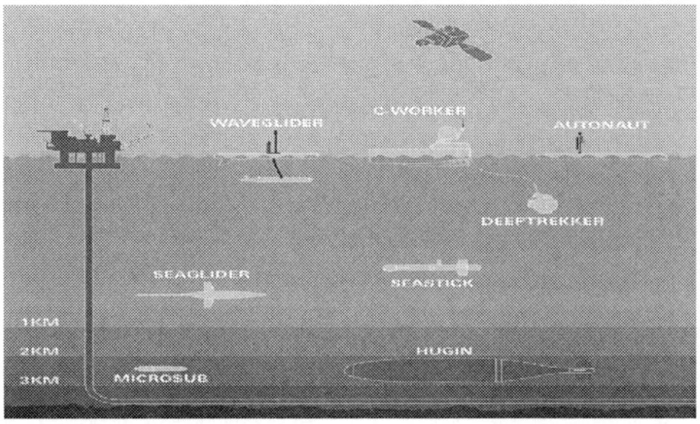

Figure 2: An example application of collection of UMS (from *"How a new robot fleet is monitoring the underwater world"* at http://www.bp.com/en/global/corporate/bp-magazine/innovations/ocean-monitoring-with-robot-technology.html)

Because not all maritime traffic (including military and commercial) always follow the COLREGS, therefore, assured autonomous navigation and control are very important to develop and maintain in USVs.

There are a wide range of definitions of autonomy depending on the field. In the maritime sector, the most widely agreed definition is given in (UK,), and has six levels. These levels are, in ascending order of autonomy:

- Level 0: Manned; Vessel/craft is controlled by operators aboard
- Level 1: Operated; Under Operated control where all cognitive functionality is within the human operator
- Level 2: Directed; Under Directed control some degree of reasoning and ability to respond is implemented into the Unmanned Vessel. However, the authority to make decisions is with the operator
- Level 3: Delegated; The Unmanned Vessel is now authorised to execute some functions. The control initiative emanates from the Unmanned Vessel and decision-making is shared between the operator and the Unmanned Vessel
- Level 4: Monitored; The Unmanned Vessel will sense environment and report its state. The operator may monitor the events, and
- Level 5: Autonomous; The Unmanned Vessel will sense environment and report its state. The operator may monitor the events.

At the very highest level, missions are specified as a series of waypoints, with functionality tags (such as profile, station keep, dock) with the vehicle attempting to maintain a straight course between waypoints. Many USVs have the capacity for real-time bidirectional communication between the control station and the USV, where the USV can have the waypoints, steering, and communication commands sent to it in near-real time.

During USV navigation, there are three main operations to carry out. These are: (a) Route Planning (waypoints' elicitation), (b) Monitoring of the Navigation, and (c) Updates to Route Planning, if and when necessary. Once the USV starts moving, locomotion along the route has to be monitored, by the USV and/or the control station, for various reasons, such as obstacle avoidance, asset detection, and changes in weather, thereby making accurate and secure navigation of these autonomous vessels essential for safety.

2.1 Navigating Autonomous Marine Vessels Safely and Securely

USVs are required to be at least as safe as the equivalent human-operated surface vessels. Some of the safety concerns for USVs that have navigation as core include: (a) their ability to avoid collisions with other marine assets, such as floating objects (e.g. bouys, etc.) or other marine vessels, (b) their ability to navigate safely in coastal areas, (c) ability to handle emergencies, such as failure recovery and repairs at sea of itself or of other marine vessels.

To be safe in its operation, a USV should endeavour not be a safety hazard to itself, other surrounding marine assets, or the maritime environment, of which it is a part. USV navigation is usually provided by the Global

Navigation Satellite Systems (GNSS), of which the General Positioning System (GPS) is a part of. Depending on the level of autonomy of the corresponding USV, successful navigation, and mission operations, require precise positioning, timing and collision avoidance. All these depend on the accuracy of the GNSS values provided to the USV.

2.2 Positioning And Navigating with Global Navigation Satellite Systems (GNSS)

Global Navigation Satellite Systems (GNSS), such as the Global Positioning System (GPS), Russia's GLONASS, the European Union's Galileo and China's COMPASS, provide important positioning, navigation and timing information to military, civilian and commercial users around the world. GNSS comprises mainly three components (Ioannides et al., 2016): (a) the User Segment, (b) the Control and Uplink Segment and (c) the Space Segment.

The Space segment consists of a constellation of operating satellites that transmit one-way signals that give the current GNSS satellite position and time. These signals are generated by the satellites' payloads that also contain one or more atomic clocks. These clocks are used to precisely time the signals and to provide good frequency reference. The navigation signals are optimised for various applications but they share a similar structure. For a given satellite, m, the transmitted signal, $s(t)$, is modelled by: $s_m(t) = \sqrt{2P_m}c_m(t)cos(2\pi f_{RF}t)$, where m denotes the satellite index, P is the transmit power, $d(t)$ the broadcast navigation message, $c(t)$ a pseudo-randomly alternating chipping sequence, t denotes time, and f_{RF} is the nominal carrier frequency.

The control segment consists of a global network of ground facilities that track the satellites, monitor their transmissions, perform analyses, and send commands and data to the constellation. As the locations of these stations are precisely known and the orbital motion of the satellites follows Kepler's laws, these data can be used to determine and predict the satellite positions. The user segment consists of the GNSS receiver equipment, which receives the signals from the GNSS satellites and uses the transmitted information to calculate the user's three-dimensional position, velocity and time. Figure 3 shows a schematic diagram of GPS showing the three segments.

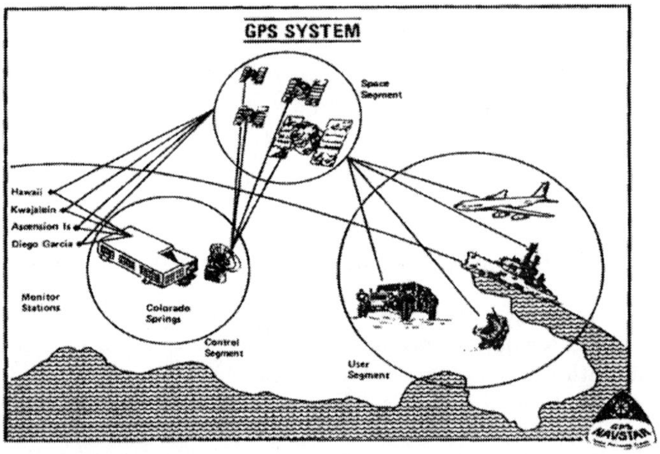

Figure 3: The Three GPS Segments, from (Humphreys, 2011)

2.3 GNSS Vulnerabilities

GNSS signals are very weak, as low as $-160dBW$, and unencrypted. As such, the system is vulnerable to unintentional and intentional interference. The result of such interference could be the complete failure of the vessel's GNSS receiver or, possibly worse, the presentation to the vehicle of hazardously misleading information for navigation and situational awareness. In general, three attack types are distinguished (Maarse, 2016): spoofing, jamming, and meaconing attacks. Meaconing is the interception and rebroadcasting of navigation signals in order to confuse navigation, while jamming is the intentional interference of the GNSS signals via the emission of radio frequency energy of sufficient power and with the proper characteristics to prevent receivers in the target area from tracking the GNSS signals. Spoofing is the broadcast of false signals with the intent that the victim receiver will misinterpret them as authentic signals. The victim might deduce a false position fix, a false clock offset, or both.

Although GNSS signal jamming has been popular in recent times, interest in GNSS spoofing has intensified of late due to successful "spoofing in the wild" that have been reported. Examples include, the Iranian military forcibly capturing a highly classified CIA drone in Dec. 2011. An Iranian engineer involved in the capture claimed that they spoofed the drone into landing in Iran when it thought it was landing at its base in Afghanistan (Rawnsley,). A scientific satellite was reported to have received spoofing-like GPS interference over Ukraine (Divis, 5 09). A yacht was spoofed deluding the receiver, causing the vessel's autopilot system and crew to navigate along a course laid out by the adversary (Bhatti and Humphreys, 7 05; ?). Spoofing could send marine vehicles, off-course, especially in low-visibility conditions, threatening safety and security.

2.4 Safe and Secure Navigation of Unmanned Marine Systems (UMS)

The rising level of autonomy brings with it new hazards and risks that need to be handled and/or mitigated in order to enjoy its economic benefits. Due to their importance, safety and security impose constraining requirements that need to be fulfilled in the design and implementation of the navigation module of unmanned marine systems, such as USVs. Safety analysis methods, such as HAZOP (Tyler et al., 2015), work on an existing design and are ill-suited to assess the kinds of cyber-physical systems employed in UMS. Systems and system designs have become so complex that waiting until a design is completed to perform safety and security analyses on it is impractical. Even if by dint of sheer will it is possible to perform such analyses, changing the design after the fact is usually impractical (financially and intellectually). Much of this effort, therefore, goes into proving that existing designs are safe and/or secure rather than building designs that are safe from the beginning. The only hope for practical and cost-effective safe design approaches in these systems is to design safety and security in from the beginning.

Due to the interactions between the software and the physical parts of cyber-physical systems, and the ensuing emergence that are the results of these component interactions, research suggests that designing secure safety-critical systems poses a substantial challenge (Oates et al., 2013) with a view that engineering "complex embedded and cyber-physical systems requires a holistic view on both product and process" (Schlinglof, 2016). Security and safety need to be incorporated across the engineering life-cycle to ensure such systems are safe from accidents and hazards, and secure from deliberate threats.

3 Security and Hazard Analyses of the Navigation Component of Unmanned Marine Systems (UMS)

Accidents have traditionally been conceived of as occurring from a sequence of directly related failure events, each of which leads to the next event in the chain of events. Increased system complexity and interactions, and the introduction of software, are leading to new types of accidents, accidents that are more a result of inter-component interactions (and not just intra-component failures). Traditional analysis methods work with accident models that are based on the fault-error-failure chain (Avizienis et al., 2004). While these models are valid to describe failures of single components, they are insufficient to describe system failures in complex interconnected systems. Systems-Theoretic Accident Model and Processes (STAMP) (Leveson, 2004) is an accident causality model based on systems theory. It expands the traditional model of causality beyond a chain of directly-related failure events or component failures to include more complex processes and unsafe interactions among system components.

STAMP is based on the three concepts of safety constraints, a hierarchical safety control structure and process models. STAMP considers events leading to accidents occur because safety constraints were not successfully enforced. Safety constraints on a system are imposed by the laws of physics, the regulatory and organisational frameworks, the systems with which it interacts, and/or the functions it performs, and design and development decisions. System-level constraints are first identified and responsibility for enforcing them is divided and allocated. Then, during system design and development, system-level safety constraints are broken down and sub-constraints are allocated to the system components.

STAMP considers systems as hierarchical control structures where each level imposes constraints on the activity of the level beneath it. The standard control structure involves four components of Controller, the Controlled entity, Actuators and Sensors (figure 4).

The controller issues control actions implemented by actuators that affect the state of the controlled entity/process. Sensors capture changes in the state of the controlled process and transmit them to the controller

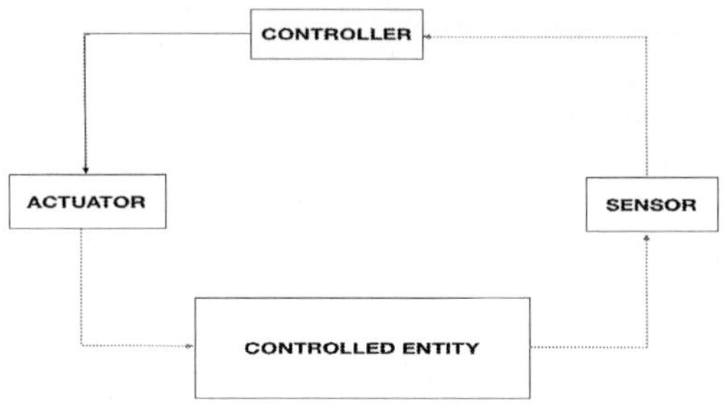

Figure 4: Standard control structure as used in STAMP

process which uses this feedback information to issue new actions to keep the controlled process in the desired safe operational state. In STAMP, the controller has a model of the controlled process. The controller maintains this model with the feedback information provided by the sensors and, based on this model, determines the control actions.

Accidents, in STAMP, are violations of safety constraints that were not adequately enforced by control actions because the model of the controlled process in the controller departs from the actual behaviour of the controlled process. This discrepancy is the source of the four possible causes of accidents: (a) A control action required for safety was not provided; (b) An unsafe control action was provided; (c) A control action required for safety was provided too early or too late or in the wrong sequence; and (d) A control action required for safety was stopped too soon or applied too long.

Based on STAMP, System Theoretic Process Analysis (STPA) (Leveson and Thomas, 2018) and STPA-Sec (Young and Leveson, 2013) were developed as new hazard analyses techniques to evaluate the safety and security of a system. STPA starts from fundamental system engineering activities, including the identification of losses or accidents to be avoided, the hazardous behaviour that could lead to these losses, safety requirements and constraints, and the basic system control structure used to avoid these losses. STPA relies on four safety related activities of system engineering, viz: (a) determination of unacceptable accidents. An accident, in STPA, is defined as "an unplanned or undesired event that result in a loss of human, human injury, property damage, environmental pollution, mission loss, etc."; (b) determination of the system boundaries. Boundaries determine which conditions related to accidents are considered part of the system and which are considered part of the environment; (c) Identification of high-level system hazards. A hazard is a system state or a set of conditions that, together with a particular set of worst-case environmental conditions, will lead to an accident; and (d) the identification, determination and definition of system safety constraints. System safety constraints are the conditions the system itself, its organisation and its development process must fulfil to prevent hazards from occurring. STPA-Sec buttresses STPA in being used to analyse the security of systems. It changes the traditional bottom-up approach to security, where threats are used to derive the security requirements, to a top-down approach where the outcomes are more relevant.

STPA has two steps: (1) the identification of potential inadequate control actions that can lead to hazardous states; and (2) the determination of how these unsafe control actions can occur. Through these two steps, STPA and STPA-Sec help to generate detailed safety and security requirements and constraints that must be implemented in the design in order to prevent the identified unacceptable losses.

3.1 Using STPA and STPA-Sec in the Security Analyses of UMS Navigation Module

Two main steps can be identified in STPA Analysis: (1) Identification of the possible accidents in the system, of system-level hazards leading to these accidents, and of the system constraints that can prevent and/or mitigate these hazards; and (2) identification of the scenarios that could lead to these unsafe control actions. Table 1 shows the results of our applying STPA Analysis on the UMS Navigation Module, helping to discern the accidents, the system-level hazards, and the constraints of the Navigation Module (NM).

3.1.1 Identification of accidents, system-level security hazards and security constraints

In table 1, we see the accidents (**AX**) that may occur in the system, the hazards that may cause the accidents to occur (**HY**), and the security constraints (safe control actions) (**CZ**) that can prevent or mitigate such hazards. Section 3.1.2, below, expands on and identifies the unsafe actions that may be caused by these hazards.

A1	Unavailability of the Navigation Module (NM)		
	H1.1	NM receives continuous stream of un-usable GNSS signals via a jamming attack	
		C1.1	NM shall assure assure the accurate and timely receipt of received GNSS signals
	H1.2	Malevolent manipulation of hardware and software layers of NM to alter GNSS data interpretation	
		C1.2	NM shall guarantee the authenticity of its software and hardware components
A2	Un-authorised disclosure of GNSS information		
	H2	There is un-authorised disclosure of GNSS information	
		C2	NM shall assure GNSS data are disclosed only to authorised parties
A3	Received GNSS information are not genuine		
	H3.1	GNSS information have been intentionally altered via a spoofing attack	
		C3.1	NM shall assure assure the integrity of received GNSS signals
	H3.2	NM receives replayed GNSS signals via a meaconing attack	
		C3.2	NM shall assure the integrity of received GNSS signals
	H3.3	GNSS information have been un-intentionally altered (probably due natural accident)	
		C3.3	NM shall assure the integrity of received GNSS signals
A4	NM physical antennae poorly installed		
	H4.1	Antennae installed position inhibits clear view of sky or clear signals from satellites	
		C4.1	NM operational staff shall assure correct installation of NM's antennae
	H4.2	Antennae improperly matched to NM receiver	
		C4.2	NM operational staff shall assure accurate matching of antennae to receiver
A5	Users' Psychological Error		
	H5	Over-reliance of users on provided GNSS data	
		C5	NM shall assure availability of other sources of accurate UMS positioning and timing data
A6	Unintentional Radio Frequency (RF) interference		
	H6	Noise from nearby RF transmitters interferes with genuine GNSS signals	
		C6	NM shall assure un-intentional RF interference

Table 1: Accidents, system-level security hazards and security constraints in Navigation Module of UMS

3.1.2 Identifying Unsafe Control Actions (UCA)

After the preliminary hazard analyses carried out in table 1, the next step is to use STAMP's four general categories of unsafe control actions (Leveson and Thomas, 2018) of: (a) "Not providing causes hazard", (b) "Providing causes hazard", (c) Wrong timing/ordering causes hazard", and "Stopping too soon/applying too long causes hazard", to identify the conditions under which the hazardous controls, as enumerated in table 1, could lead to system hazards.

Tables 2 and 3 present our use of STAMP to analyse some controls and outputs issued by the NM. The first column identifies the analysed control. The second column records the consequences of not providing a safe control. The third column records the consequences of providing an unsafe control (i.e. the controller allows the controlled process to perform actions in a context where hazards may occur). The fourth column records the consequences of providing a safe control too early or too late or in a wrong order. The fifth column records the consequences of stopping a safe control too soon or applying it too long. Every UCA must be traceable to one or more system-level hazards (Leveson and Thomas, 2018).

These identified unsafe control actions, with the related hazards, serve many functions. They can be used to shape early design decisions regarding the security of the system to be built. When the conditions under which a control action may be unsafe are stated, these help the engineers to perceive those instances, eliminate those instances from the system design or find ways to mitigate them. When translated into requirements, they form parts of the constraints to be enforced by the system design.

Control Action	**Not Providing Causes Hazard**	**Providing Causes Hazard**	**Wrong Time or Wrong Order Causes Hazard**	**Stopped Too Soon or Applied Too Long**
NM assures authenticity of its software & hardware components	**UCA1**: Malevolent components installed in NM [H1.2] **UCA2**: Unauthorised principals able to alter GNSS data interpretation [H1.2, H2, H3.2]	None	Same as UCA1 & UCA2	Same as UCA1 & UCA2 (Stopped Too Soon)
NM mitigates jamming attack	**UCA3**: Denial of Service attack (of GNSS data) [H1.1, H5]	None	Same as UCA3	Same as UCA3 (Stopped Too Soon)
NM provides authorised disclosure of GNSS data	**UCA4**: Un-authorised entities able to elicit USV position (information leakage) and/or able to replay GNSS information (traffic analyses). Possibility of taking control of USV & sabotage mission [H2, H3.1, H3.2] **UCA5** Tampering, i.e., deliberately destroying or corrupting data [H2, H5]	None	Same as UCA4	Same as UCA4 (Stopped Too Soon)
NM assures integrity of GNSS data	**UCA6**: Spoofing attack: Received GNSS data probably intentionally altered [H3.1, H5S] **UCA7** Meaconing attack: Received genuine GNSS data replayed back to USV after timed delay, leading to GNSS error readings [H3.2, H5]	None	Same as UCA6 & UCA7	Same as UCA6 & UCA7 (Stopped Too Soon)

Table 2: Unsafe Control Actions of the Navigation Module of UMS (1/2).

4 Conclusion and Future Work

The vast majority of global trade flows across the world's oceans. Unmanned marine systems (UMS) are increasingly being used to facilitate and secure these trade flows. The security of the autonomous navigation of these systems is becoming increasingly important. Therefore, accurate positioning, velocity, and timing (PVT) values are essential to their safe navigation. These PVT values are usually provided by the Global Navigation Satellite Systems (GNSS) signals that are received and decoded by the UMS' receivers. These signals are very weak and un-encrypted. Their accurate reception and decoding are, therefore, open to many vulnerabilities. This paper has used systems, and control, theory and STPA to analyse these vulnerabilities. We identified the major system-level security hazards and constraints, and especially of unsafe control actions. Our analyses can be used as a springboard to drive their mitigation and/or resolution, thereby helping to designing a more effective and secure UMS' navigation components.

In future work, we will extend these analyses with STPA causal scenarios, and using the Event-B (Abrial,

Control Action	Not Providing Causes Hazard	Providing Causes Hazard	Wrong Time or Wrong Order Causes Hazard	Stopped Too Soon or Applied Too Long
NM operational staff installs antennae properly	**UCA8**: Erroneous GNSS data received [H4.1, H4.2, H5]	None	Same as UCA8	Same as UCA8 (Stopped Too Soon)
NM provides additional Positional, Velocity, Timing (PVT) sources in addition to GNSS	**UCA9**: Probability of over-reliance of users on GNSS data [H5]	None	Same as UCA9	Same as UCA9 (Stopped Too Soon)
NM provides provides mitigation of unintentional RF interference	**UCA10**: Erroneous GNSS data received [H1.1, H5]	None	Same as UCA10	Same as UCA10 (Stopped Too Soon)

Table 3: Unsafe Control Actions of the Navigation Module of UMS (Table 2 contd.) (2/2).

2010) formalism, will develop a framework for security analysis for autonomous navigation of unmanned marine systems.

References

Abrial, J.-R. (2010). *Modeling in Event-B: System and Software Engineering.* Cambridge University Press.

Air Force Research Laboratory, Dayton, O. (2013). U.s. air force, autonomy science and technology strategy.

Avizienis, A., Laprie, J.-C., Randell, B., and Landwehr, C. (2004). Basic concepts and taxonomy of dependable and secure computing. In *IEEE Transactions on Dependable and Secure Computing*, volume 1, pages 11–33.

Bhatti, J. and Humphreys, T. (2017-05). Hostile control of ships via false gps signals: Demonstration and detection. *Navigation.*

Divis, D. A. (2015-09). Scientists document possible drone jamming. *Inside Unmanned Systems.*

Humphreys, T. (2011). State of the art and future trends in radionavigation. Briefing to USPTO.

Ioannides, R. T., Pany, T., and Gibbons, G. (2016). Known vulnerabilities of global navigation satellite systems, status, and potential mitigation techniques. In *Proceedings of the IEEE*, volume 104.

Leveson, N. (2004). A new accident model for engineering safer systems. *Safety Science*, 42(4):237–270.

Leveson, N. G. and Thomas, J. P. (2018). *STPA Handbook.*

Maarse, M. (2016). A systematic approach towards gnss receiver vulnerability analysis on remotely piloted aircraft systems.

Navy, U. S. United states navy biography. http://www.navy.mil/navydata/leadership/quotes.asp?q=253&c=6. 2018-06-28.

Oates, R., Thom, F., and Herries, G. (2013). Security-aware, model-based systems engineering with sysml. In *Proceedings of the 1st International Symposium on ICS & SCADA Cyber Security Research*, pages 78–87.

Rawnsley, A. Iran's alleged drone hack: Tough, but possible. http://www.wired.com/dangerroom/2011/12/iran-drone-hack-gps.

Rolls-Royce. Rolls-royce written evidence (auv0083). http://data.parliament.uk/writtenevidence/committeeevidence.svc/e and-technology-committee-lords/autonomous-vehicles/written/42075.html.

Schlinglof, B.-H. (2016). Cyber-physical systems engineering. In Liu, Z. and Zhang, Z., editors, *Engineering Trustworthy Software Systems*, volume 9506, pages 256–289. Lecture Notes in Computer Science.

Tyler, B., Crawley, F., and Preston, M. (2015). *HAZOP: Guide to Best Practice (3rd ed.)*. IChemE.

UK, M. Being a responsible industry - an industry code of practice. https://www.maritimeuk.org/documents/197/CODE_OF_PRACTICE_V1.0_-_Up_to_24m_-_Final.pdf.

Young, W. and Leveson, N. (2013). Systems thinking for safety and security. In *Proceedings of the 29th Annual Computer Security Applications Conference*, pages 1–8.

ASVTrafficSim: A simulator for Autonomous Surface Vehicle and Manned Vessel Collisions

Colin Sauze
Department of Computer Science
Aberystwyth University
Penglais
Aberystwyth
Ceredigion
United Kingdom
SY23 3DB
cos@aber.ac.uk

Abstract

At present there is only limited data about the probability of collisions between autonomous surface vehicles (ASVs) and manned vessels. This paper describes a simulator for calculating the probability of collisions between them. It is intended to help formulate safety advice and policy on where and when ASVs can and can't be safely used. The simulator tests hypothetical courses of the ASV against real traffic data recorded from automatic identification system (AIS) transponders. This allows simulated missions to be tested against real world traffic patterns. The simulator has successfully simulated example missions using a small set of example AIS data based in San Francisco bay. Future work will involve running simulations from large AIS datasets and with a variety of ASV types, as well as covering differing weather scenarios.

1 Introduction

This paper discusses the design and implementation of an Open Source software tool: ASVTrafficSim for calculating the probability of an ASV and manned vessel colliding or having a near miss. It is intended to give ASV operators a tool to evaluate the risk of potential missions and to be able to adjust their mission to reduce this risk. Recent experience from the Microtransat Challenge has been a number of ASVs accidentally retrieved by fishing boats. In several of these incidents the ASVs owners unknowingly planned courses which sailed through fishing grounds. The ability to test their routes for safety in advance might have prevented some of these incidents.

ASVTrafficSim uses data recorded from Automatic Identification System (AIS) transponders to compare a simulated course with data from real traffic. By running multiple iterations of ASVTrafficSim each with slightly different courses accurate probabilities of collisions can be determined. It is also possible to determine which times of day, days of the week or times of year might be safer when comparing against regular traffic such as passenger ferries.

In: S. M. Schillai, N. Townsend (eds.): Proceedings of the International Robotic Sailing Conference 2018, Southampton, United Kingdom, 31-08-2018

1.1 AIS

AIS transponders broadcast the boat's maritime mobile service identity (MMSI), position, heading, speed, rate of turn every few seconds over a VHF radio link. AIS is required by all manned vessels over 300 gross tonnes, with many smaller vessels also opting to use the system. Less frequently sent messages can also include the ship's destination, name and dimensions. Two classes of AIS transmitter are available, class A and class B. Class A is intended for commercial traffic and transmits at higher power (12.5W) and more frequently, its typical range is around 40 nautical miles (Marine Management Organisation, 2014). The lower power class B system is intended for pleasure craft and only uses 2 watts of transmission power and transmits less frequently, its typical range is around 10 nautical miles(Marine Management Organisation, 2014). This lower power and frequency combined with many smaller boats still not being equipped with AIS mean that data for small craft is often lacking.

A number of internet linked shore receivers pass on data to web services such as MarineTraffic (Marine Traffic, 2018) and AISHub (AISHub, 2018). In recent years satellite based reception has also become increasingly common, allowing sites like MarineTraffic to cover traffic out of range from shore. Analysis of shipping patterns from AIS data has become increasingly common (Ristic et al., 2008) (Fiorini et al., 2016). One common output of these analyses is traffic density maps which show a generalised view of how busy an area is. MarineTraffic(Marine Traffic, 2018), the US Coastguard(Beuaru of Ocean Energy Management, 2018) and UK government(Marine Management Organisation, 2014) are amongst those publishing density maps. Although these density maps can form a useful tool in planning ASV operations they are relatively crude and often lack time of day, day of week or time of year specific information. ASVTrafficSim aims to provide more a detailed and specific view of the risks involved with travelling a given route.

1.2 AIS Data Sources

To test against real traffic patterns a source of AIS data was needed. There are a number of commercial sources of AIS data, however the cost of obtaining raw data from these was prohibitively expensive and not an option for this research. Several free alternatives were identified. The Marine Cadstre (Beuaru of Ocean Energy Management, 2018) from the US Coastguard offers a large amount of AIS data, but this is formatted for the commercial ArcGIS software and is difficult to read without ArcGIS. Several volunteer run networks make data freely available online. These include AISHub (AISHub, 2018) and aprs.fi. The data on aprs.fi was found to be lacking in coverage and there don't appear to be many AIS receivers in their network. AISHub operates an exchange where free access to data is given in return for contributing data. At time of writing the author has not setup a receiver capable of contributing to AISHub, but intends to do so in future. The Exploratorium Museum in San Francisco operates a receiver on the San Francisco seafront that continuously streams raw data to their website (ais, a). Greenpeace have a collection of 1.2 million messages taken over two days in 2014 is available from their github page (ais, b). This data is in raw NMEA format and can be decoded using the open source LibAIS (Schwehr, 2015) software.

2 ASVTrafficSim

ASVTrafficSim is an open source tool written in Python, it is available from (Sauze, 2018). It simulates sailing of a target route using the open source Sailsd simulator(Taylor, 2016) which has a reasonably accurate physics model based upon work by (Jaulin and Le Bars, 2012). The autopilot logic for controlling Sailsd operates through Boatd, another open source tool for presenting an HTTP and JSON based interface to an ASV. Output from Boatd is sent over a UDP datastream in the form of NMEA0183 GPS strings, these are received and processed by ASVTrafficSim's collision detector which uses them to establish the ASV's current location. Upon starting ASVTrafficSim loads a datafile of AIS messages and parses these with LibAIS, extracting each report's Maritime Mobile Service Identity (MMSI), ship name (where given), latitude, longitude and time. This data was then used to test for collisions at the simulated time and location. Figure 1 shows how these components are integrated and communicate with each other.

AIS data is interpolated between data points on a 1 second basis. Typically class A AIS data is transmitted every few seconds, so there is relatively little error in the interpolation. However class B messages are less frequent, typically being 10s of seconds apart and due to the lower power they are often not received as easily by shore stations. The linear interpolation was limited to a maximum of 15 minutes so that boats which are moored and have switched off their AIS aren't projected as continuing to travel. This linear interpolation method will not be completely accurate as it only uses the last heading and speed and ignores if the boat was rotating. A small improvement to the accuracy of class B interpolation might also be possible using polynomial interpolation instead of linear interpolation.

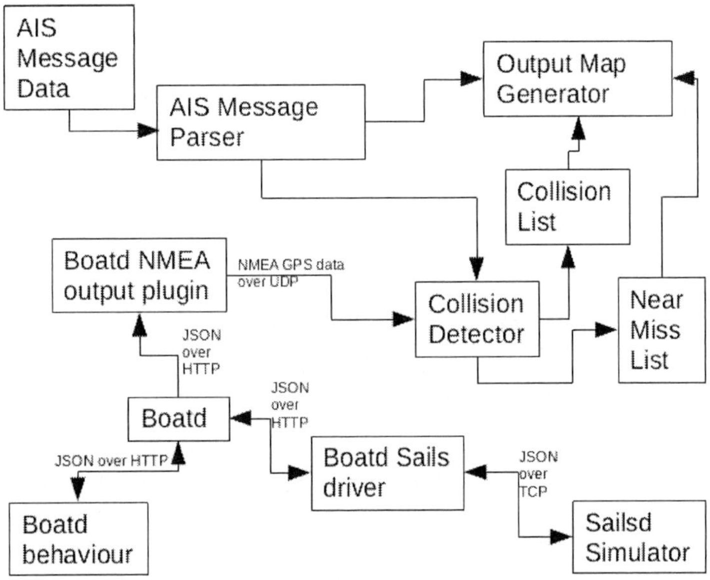

Figure 1: The data flows between the components in ASVTrafficSim and its dependencies.

The Sailsd simulator operates on a realtime basis, to maintain realism ASVTrafficSim also does this. Each second that time advances in Sailsd corresponds to one second in AIS data of ship movements. This does have the disadvantage of simulations taking a long time to run. Ship movements are linearly interpolated between data points in the AIS data and these are generated on a second by second basis. At each one second iteration the position of every ship is compared with the position of the simulated ASV. Ideally collisions could be calculated by looking at the ship's dimensions and working out if it overlapped with the ASV. Unfortunately the Exploratorium AIS dataset is missing any messages covering boat dimensions. Therefore any ship getting within 10 metres of the ASV is considered to be a collision and 100 metres a near miss. All collisions and near misses are recorded and saved by the output map generator. This saves data into a GPX file along with the ship and ASV tracks.

3 Results

A simulated course was setup to sail a square around Alcatraz Island in San Francisco harbour, the simulation was set to use a northerly wind requiring upwind tacking on the northbound leg of the course. This crosses busy shipping channels in and out of San Francisco bay, tourist boat traffic going to and around the island and ferries operating from the city centre. The course taken on this route is shown in figure 2. The 2014 Exploratorium dataset from the Greenpeace was used to supply traffic information. The simulation run took place between 12:40 and 14:53 UTC (4:40 and 6:53 PST) on November 26th 2014. These times represent the start of the AIS dataset. 10 laps of the course were completed in this time.

In running this simulation for just over two hours, there were two collisions and 69 near misses. All of these were with the ferry Zelinsky which operates tours around the island. Both of the collisions occurred when the boats were less than 10 metres apart over the course of two seconds. The near misses are clustered into two areas, one on the western side of the course where both boats were sailing in parallel and very closely. The others are either side of the point where the collision occurred. Figure 3 shows a map of this output, with the near misses marked as triangles and the collisions as circles.

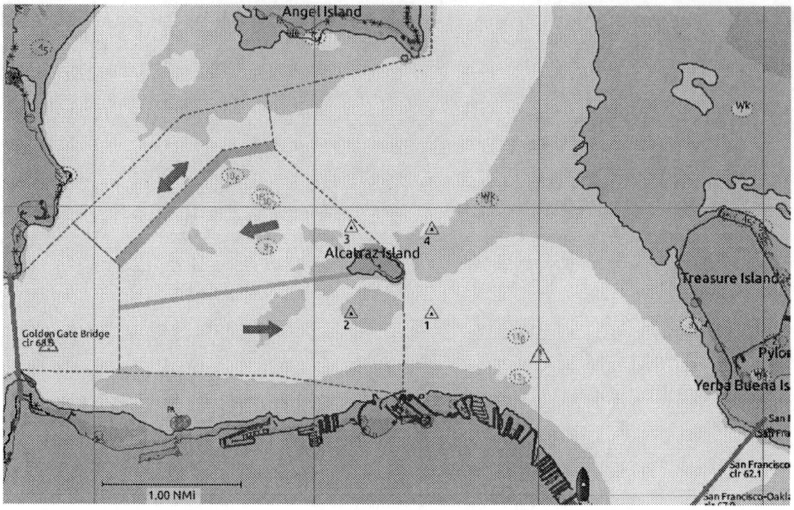

Figure 2: A map showing the waypoints used in the test route. Each waypoint is indicated with a triangle. Map from NOAA ENC Chart US5CA13M (National Oceanic and Atmospheric Administration, 2018).

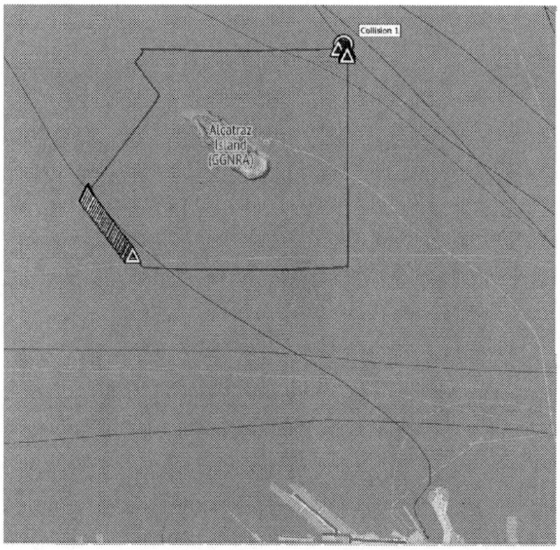

Figure 3: A map showing the collisions and near misses between the simulated ASV and real traffic. Each near miss is indicated with a triangle and a collision with a circle. The two collisions are within a few metres of each other and the first is obscured by the second in this map. The red line around the island is the ASV's track and the dark blue line is from the ferry Zelinsky. The other coloured tracks are from other vessels. Map tiles from Stamen, licensed under CC BY 3.0. Map data from Open Street Map contributors, licensed under CC BY-SA.

4 Conclusion

This work demonstrates the viability of ASVTrafficSim for detecting potential collisions. It allows fine grain identification of which vessels a collision occurred with or came close to occurring with. This is particularly valuable when planning routes in areas with regular traffic such as ferries. By running multiple simulation runs with differing start times or route predictions over wider areas can also be obtained.

5 Future Work

Future work will involve improving the usability and accuracy of ASVTrafficSim. Usability could be improved so that a user can simply input the parameters of their USV and intended course and then receive a report showing the collisions and near misses. Currently the user must start five different programs to run the simulation, although Docker and Singularity containers are available to simplify this. Run times could be reduced by removing the real time nature of the simulator and running at faster than real time. At present the physics model available within Sailsd is still somewhat unrealistic. More fine tuning of its parameters is required to match the performance of a real USV. It does not include the effects of sea state, tides, currents or simulate any variations in the wind. A more realistic simulation could be achieved by including these variables. Tides are of particular importance given how small ASVs are often unable to travel fast enough to fight against strong tides. To give a clearer idea of the probability of a collision a monte carlo method could be used with the simulation being repeated many times with small random variations in weather conditions or the course being sailed.

The use of larger AIS datasets will also help to improve the range of simulations which can be run and the geographic areas which can be tested. Obtaining AIS data remains an obstacle, although commercial providers of such data exist their prices are prohibitively expensive. AISHub (AISHub, 2018) is a potential alternative to this as they will provide access to data from their network in exchange for contributing a receiver. Alternatively datasets could be generated by setting up receivers comparable to the one used by the Exploratorium.

Acknowledgements

The author would like to thank Louis Taylor with his help with getting Boatd and Sailsd to work with this project. We acknowledge the support of the Supercomputing Wales project, which is part-funded by the European Regional Development Fund (ERDF) via Welsh Government.

References

AIS Exploratorium. `http://ais.exploratorium.edu` (Accessed Aug 2nd 2018).

AIS Exploratorium Data - November 26th-27th 2014. `https://github.com/greenpeace/oceansofdata/tree/master/ais-exploratorium-edu` (Accessed Aug 2nd 2018).

AISHub (2018). AIS Data Exchange. `http://www.aishub.net` (Accessed Aug 2nd 2018).

Beuaru of Ocean Energy Management (2018). Marine cadastre national viewer. `https://marinecadastre.gov/nationalviewer/` (Accessed Aug 2nd 2018).

Fiorini, M., Capata, A., and Bloisi, D. D. (2016). AIS data visualization for maritime spatial planning (msp). *International Journal of e-Navigation and Maritime Economy*, 5:45 – 60.

Jaulin, L. and Le Bars, F. (2012). A simple controller for line following of sailboats. In *proceedings of the 5th International Robotic Sailing Conference, Cardiff, United Kingdom, September 2012*, pages 117–129.

Marine Management Organisation (2014). Mapping UK shipping density and routes from AIS.

Marine Traffic (2018). Global ship tracking intelligence. `http://www.marinetraffic.com` (Accessed Aug 2nd 2018).

National Oceanic and Atmospheric Administration (2018). ENC chart US5CA13M: San francisco bay candlestick point to angel island. `http://www.charts.noaa.gov/ENCs/Agreement.shtml?US5CA13M` (Accessed Aug 2nd 2018).

Ristic, B., Scala, B. L., Morelande, M., and Gordon, N. (2008). Statistical analysis of motion patterns in AIS data: Anomaly detection and motion prediction. In *2008 11th International Conference on Information Fusion*, pages 1–7.

Sauze, C. (2018). ASVTrafficSim. `https://github.com/colinsauze/ASVTrafficSim` (Accessed Aug 2nd 2018).

Schwehr, K. (2015). LibAIS. `https://github.com/schwehr/libais` (Accessed Aug 2nd 2018).

Taylor, L. (2016). Sails simulator: A set of tools for robotic sailing boat simulation. `https://github.com/sails-simulator` (Accessed Aug 2nd 2018).

Part III

Manoeuvre & Route Planning

Reliability informed routing for Autonomous Sailing Craft

Thomas Dickson
Fluid Structure Interactions
B176, Boldrewood Campus, S016 7QF
thomas.dickson@soton.ac.uk

James Blake
Fluid Structure Interactions
B176, Boldrewood Campus, S016 7QF
J.I.R.Blake@soton.ac.uk

David Sear
Geography and Environment
Highfield Campus, SO17 1BJ
D.Sear@soton.ac.uk

University of Southampton

Abstract

This paper introduces a novel method for modelling the influence of likely autonomous sailing craft failure conditions into the route planning algorithm. The accuracy of the original route planning algorithm is quantified using numerical error estimation techniques. It was found that over the course of a Trans-Atlantic voyage a grid size of 36 km produced an error of ± 2.1 hours over the course of a 703.25 hr voyage. The implementation of the failure model within the routing algorithm is verified using a control weather scenario. This verification is shown to be significant with respect to the method's numerical error. Future work will involve gathering evidence on failure criteria in order to update the failure model.

1 Introduction

The Microtransat challenge is a competition undertaken by autonomous sailing craft (ASC) to complete an Atlantic crossing in the fastest time (Microtransat Challenge, 2018). To the authors knowledge, the Microtransat challenge has not been successfully completed as a consequence of competing vehicles failing before they complete the voyage. This paper seeks to address this problem through introducing a voyage failure model into a route planning algorithm, thereby identifying a route that manages the risk of failure. One use of this method is to assist with operational route planning with an existing craft. Another use could be to assist with the design process through modelling the influence of potential failure mechanisms.

Autonomous sailing craft are sailing robots which are designed to operate independently of human control or maintenance after their planned voyage has started. This independent operation requires that the reliability of the entire system must be extremely high. This reliability must be modelled based on the environmental conditions that are likely to be experienced. It is likely that through considering the reliability over the range

In: S. M. Schillai, N. Townsend (eds.): Proceedings of the International Robotic Sailing Conference 2018, Southampton, United Kingdom, 31-08-2018

of different environmental conditions experienced on a Trans-Atlantic voyage it will be possible to improve the likelihood of completing a voyage successfully.

The East-West Trans-Atlantic voyage between the Bay of Biscay and the Caribbean is one of the most travelled maritime voyages in history. Consequently the common environmental conditions experienced are well known. Research into the navigational challenges experienced for both directions of the MicroTransat competition (East-West, West-East) has identified that the only route that compares with the West-East leaves from the Canaries and not the current start location of the MicroTransat challenge (Schlaefer and Blaurock, 2011). (Schlaefer and Blaurock, 2011) presented analysis that used failure frequency to estimate the likelihood of completeing a voyage, although wasn't able to relate this to the environmental conditions encountered. The East-West Microtransat route appears to be the most challenging route as participating ASCs fail earlier on this voyage relative to the other direction (Microtransat Challenge, 2018). Through modelling different failure modes within the routing algorithm it will be possible to understand why failure occurs earlier on the East-West route and to improve the design process.

Reliability engineering is the analysis of the different possible failure modes of an engineering artefact. Many different methodologies exist to achieve this. A Fault Tree was used to model the different factors which may cause a structural failure for a sailing craft (Auboin, L., Blake, J. I. R. and Turnock, 2010). A failure mode and effect analysis (FMEA) using fuzzy based failure modes was shown to be effectively applied to a yacht fire system design (Helvacioglu and Ozen, 2014). This paper demonstrated how it was possible to flexibly model expert opinion on the efficacy of a system through ranking hazards and then using fuzzy logic to model the influence of the lack of knowledge.

Bayesian belief networks (BBNs) have been used to model a complex network of empirical data and expert opinion which calculated the probability of Autonomous Underwater Vehicle failure given a specific voyage scenario (Brito and Griffiths, 2016). Various different reliability engineering and design problems have been shown to be capably modelled using BBNs (Friis-Hansen, 2000). Given the lack of data on sailing craft design the flexibility to incorporate different failure mechanisms and sources of information mean that BBNs offer the most suitable framework to model the reliability of an ASC. The output of a BBN is a probability of some event occuring given a range of input factors which can be structural or environmental.

Sailing craft route planning uses environmental factors to determine the optimal route for a sailing craft to take given its performance and the specific optimisation algorithm used. The most important environmental factor considered is the wind vector, this can be generated using a wind model, reanalysis data or weather forecast. The first research into solving the sailing craft route planning problem used a recursive dynamic programming formulation which divided the domain into nodes over which the shortest path was calculated (Philpott and Mason, 2001). Different wind models (Philpott et al., 2004; Dalang et al., 2014) or race strategy and opponent models (Spenkuch, 2014; Tagliaferri and Viola, 2017) have been used to improve the accuracy of sailing routing models.

The influence of different methods of modelling the ability for a sailing craft to sail upwind has been explored (Stelzer and Pröll, 2008) along with modelling the time taken to complete course changes (Ladany and Levi, 2017). The chief drawback of these methods is that they become unwieldy when applied to the long course route modelling problem. Full scale sailing craft race modelling has typically minimised either the time taken to complete a course (Ferretti and Festa, 2018), or the risk of losing to an opponent (Tagliaferri et al., 2014). To the authors knowledge, the consideration of reliability as a constraining factor in the sailing craft routing algorithm is a novel one.

This paper introduces a novel method for modelling the reliability of the sailing craft within the cost function used in the routing algorithm. This will allow the identification of routes which meet different sets of failure criteria.

2 Method

2.1 Voyage failure model

The performance of the ASC can be specified in terms of its speed and failure mechanisms. The failure mechanisms may be estimated from structural analysis or empirical data on past failures. If no information exists then it is possible to specify different combination of environmental parameters which are likely to cause failure. Using these parameters it will be possible to avoid environmental conditions which are likely to cause failure.

Figure 1 illustrates an example of a BBN which relates environmental parameters describing the wind and the waves to the calculation of the probability of voyage failure. The top rank of nodes take the values of the environmental parameters as inputs. The middle rank accept the inputs and calculate the probability of craft failure if neither parameter will cause a failure, a single parameter causes failure or both will cause failure. It is possible to model three different failure levels in this manner, although due to the flexibility of the BBN it is possible to integrate different physical models into the calculation of failure probability.

Bayesian belief networks use Bayes rule, Equation 1, in order to relate the probability of a specific event occuring given that other events have already occured. $P(A|B)$ is the probability of event A occuring given that event B is true, $P(B|A)$ is the likelihood of B occuring given that A is true and $P(A), P(B)$ are the likelihoods of A and B occuring independently of each other.

$$P(A|B) = \frac{P(B|A)P(A)}{P(B)} \tag{1}$$

The failure model used in this paper relates two failure conditions based on the wave height and apparent direction to the probability of voyage failure. The joint probability distribution can be defined as being $P(F, WD, WH) = P(F|WD, WH)P(WD)P(WH)$. The names of the model have been abbreviated as $F =$ voyage failure (true/false), $WD =$ apparent wave direction failure (true/false) and $WH =$ wave height (true/false). Note that it is possible to use continuous distributions to model events rather than binary criteria.

If the criteria for the apparent wave direction failure are met then it is possible to calculate the likelihood of voyage failure using Equation 2. The weather conditions at a specific point determine whether the failure criteria are triggered. Table 1 defines the conditional probability table which enables the calculation of the probability of failure as a consequence of a single or either wave failure criteria being met. Note that it is possible to use continuous probability distributions within a BBN which would allow more realistic modelling of reliability.

$$P(F = T|WD = T) = \frac{P(F = T, WD = T)}{P(F = T)} = \frac{\sum_{WH \in T,F} P(F = T, WH, WD = T)}{\sum_{WD,WH \in T,F} P(WD, WH, F = T)} \tag{2}$$

WD Criteria	WD Criteria	Pass	Fail
T	F	0.1	0.9
F	T	0.1	0.9
F	F	1.0	0.0
T	T	0.0	1.0

Table 1: Conditional probability table relating failure criteria combination to the probability of failure.

2.2 Route modelling

The sailing craft routing algorithm used to identify the shortest path is the recursive dynamic program formulation introduced in (Philpott and Mason, 2001). The advantage of using the dynamic programming paradigm is that it is able to guarantee that the identified route is the best candidate of all possible routes in the domain, according to Bellmans principle of optimality (Bellman, 1957). The chief drawback is the large computational cost incurred through checking all possible solutions.

To begin the process a great circle route is drawn between the start and finish location with the extents of the rectangular domain limited by the number of nodes in a rank and the distance between each node. The continuous domain is divided into discrete nodes as is illustrated in Figure 2. Each node is connected to all the nodes in the preceding and anteceding ranks to form a digraph connecting the start and the finish node.

For the position at any given node i the travel time between nodes i and a node on the next rank j along the arc (i, j) starting at time t is $c_{arc}(i, j, t) = c_{seg}(\mathbf{x}_i, \mathbf{x}_j, t)$. This function considers the wind speed and direction and course direction and calculates boat speed from known performance data. The wind speed and direction is a function of time and space and is retrieved from the wind scenario being used. $c_{seg}(\mathbf{x}_i, \mathbf{x}_j, t)$ is the time taken to sail the great circle joining location $\mathbf{x}_i$ to location $\mathbf{x}_j$ starting at time t.

The minimum time path is identified using a forward looking recursive algorithm which is described in Equation 3. $f^*(i, t)$ is the time taken for the optimal sequence of decisions from the node-time pair (i, t) to the finish node

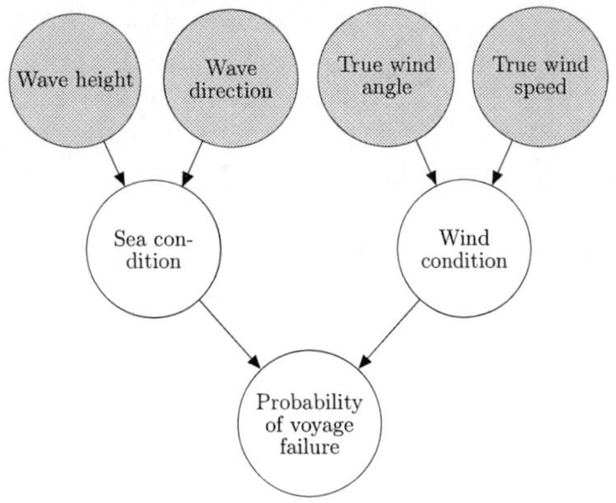

Figure 1: A Bayesian Belief Network to model the relationship between environmental conditions and the probability of voyage failure.

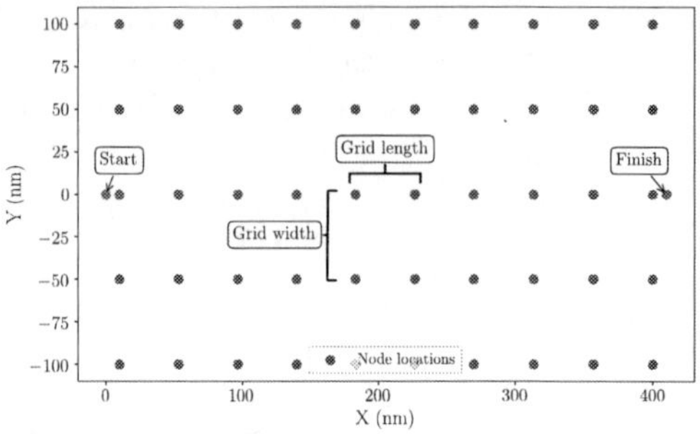

Figure 2: The continuous domain between the start and finish is discretized into points as a function of the distance between each point.

and $j^*(i,t)$ is the successor of i on the optimal path when in state t. Γ_i is the set of all nodes on the graph.

$$f^*(i,t) = \begin{cases} 0, & i = n_{finish} \\ \min_{j \in \Gamma_i} [c_{arc}(i,j,t) + f^*(j, t + c_{arc}(i,j,t))], & \text{otherwise} \end{cases}$$
$$j^*(i,j) = \arg\min_{j \in \Gamma_i} [c_{arc}(i,j,t) + f^*(j, t + c_{a}rc(i,j,t))], i \neq n_{finish} \tag{3}$$

The failure model is introduced into the c_{arc} function, as shown in Algorithm 1. Extra variables such as the weather scenario and the acceptable probability of failure ap_f are introduced. V_S is the speed for the particular set of weather conditions. If the failure model for a given segment at a given time exceeds a specified acceptable probability of failure then the time taken to complete the arc is set as infinite, otherwise the speed is interpolated from known performance data. When the shortest path is identified it naturally avoids these arcs and thus returns a route which is the minimum time route that meets the reliability requirement.

Algorithm 1 Cost function

```
function c_seg( x_i, x_j, t, ap_f, Weather scenario)
    WH_i, WD_i, TWA_i, TWS_i ← Weather scenario            ▷ Weather conditions
    p_f ← BBN Failure(WH_i, WD_i, TWA_i, TWS_i, t)
    if p_f < ap_f then
        VT = Distance_{c_seg} / V_S(WH_i, WD_i, TWA_i, TWS_i)
    else
        VT = inf
    end if
end function
```

3 Application

To verify whether the failure model is sensitive to the inclusion of failure the error generated as a result of the discretization of the domain must be calculated. The approach taken to calculate discretization error involves solving the shortest path for a control weather scenario over a range of grid sizes.

The simulation error estimated is used to inform the simulations used to demonstrate the efficacy of the failure model. In order to demonstrate the ability of the failure model, known weather conditions which trigger failure are modelled in the control weather scenario. Through specifying two different known failure conditions it will be possible to show whether and how the failure model informs the choice of route so as to avoid areas which will trigger failure. Code has been developed to support this analysis and a link is included in the acknowledgements.

3.1 Solution accuracy

The routing algorithm approximates the minimum voyaging time based on a discretization of a continuous domain. Consequently there is an associated discretization error which must be quantified in order to give credibility to the results. The route simulated is the East-West MicroTransat route using the performance of an ASV, the Maribot Vane, to give context to the results. Although the Maribot Vane does not meet the rules of the MicroTransat challenge it is a similar craft type has has experimentally validated performance data (Dhomè, 2017). The wind condition across the domain has been set at 15 knots from the North. This isolates the source of any variation in voyaging time to the variation in grid size.

To quantify the discretization error the Grid Convergence Index (GCI) was calculated (Roache, 1997; Celik et al., 2008). The GCI index calculates the difference between the estimated result and the extrapolated result calculated as a function of the trend of the previous grid sizes. This index is often used in Computational Fluid Dynamics to calculate a 95% confidence region. As it has previously not been used to interpret routing algorithm results this index can only be used to guide the interpretation of whether the grid size used is fine enough for purpose.

The grid size may be limited as soon as the voyaging time results have started to converge asymptotically as there will be little improvement in accuracy at the expense of large computational cost. Another limiting factor is the physical implication of the area enclosed in the cell, if the cell is too large it may not physically relate to the routing problem.

The results for routing simulations over a range of grid sizes can be seen in Figure 3. The control weather scenario used modelled the wind as being 15 knots from the North across the whole domain. The start location was 45° N and 12° W and the finish location was 17.5^o N and 60.0 W. Relative to the scale of the simulation the error associated with a normalised grid height of 36327.16 is 0.298%, corresponding to ± 2.1 hours. Over the course of route lasting roughly 870 hours an error of ± 2.1 hours is acceptable.

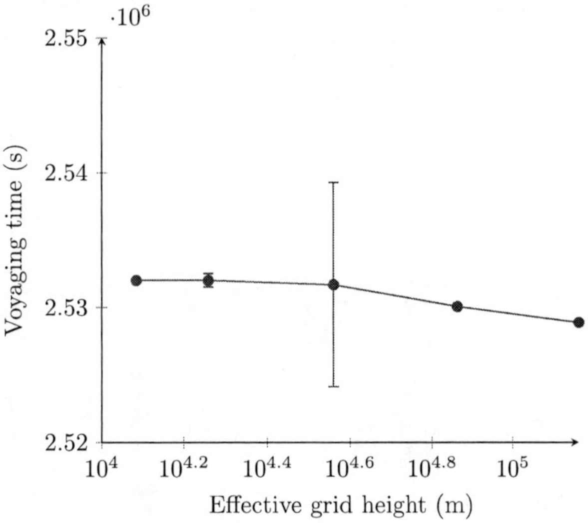

Figure 3: Grid discretization size study

3.2 Influence of failure model on routing time

This section demonstrates the application of the novel reliability routing method using a control weather scenario. This weather scenario has been modified with patches of weather designed to activate the failure model. The control weather scenario encompasses the entire routing domain and has weather parameters which are independent of time. The simulations were run using a grid spacing of 36 km which corresponds to 160 nodes along a given edge, 25600 in total.

Two rectangular patches have been introduced into the domain with specific environmental conditions designed to trigger one and then both failure criteria. Area 2 is the larger patch with the co-ordinates $\pm 5^o$ about the central point at 40° W 33° N and has the wave height parameter set to 4m and is designed to provoke the single failure criteria. Area 1 has the wave direction set to 240 degrees, approximately the reciprocal bearing of the

East-West Trans-Atlantic course and is a rectangular area with the points $\pm 3^o$ about the central point at 40° W 33° N. The combination of Areas 1 and 2 will provoke the double criteria failure model. The hypothesis is that without a failure model the shortest path will cross both areas, a double failure model will skirt Area 1 and the single failure model will skirt Area 2.

A challenge regarding the modelling of ASC failure is the lack of available data on their failure and potential failure modes, this is likely a consequence of the impracticality of their recovery on the event of voyage failure. A failure model is constructed using a BBN where the probability of failure is a consequence of different combination of environmental parameters being exceeded.

The failure model implemented has three discrete levels of failure. The initial level represents no failure criteria being exceeded, the second level represents a single failure criteria being exceeded and the final level represents routing despite any combination of failure criteria being exceeded. This model uses the structure of the BBN illustrated in Figure 1.

Two wave statistics are assumed to be significant with regards to causing the failure of an ASC, the wave direction and the wave height. For this control weather scenario the triggering failure criteria are when the wave height exceeds 4m and the apparent wave direction is under 60°. These parameters are selected in order to illustrate the ability for the BBN to avoid failure conditions rather than from any real performance data. Ship design uses mean wave height as part of the design process to estimate the motions that the ship will have to. Sailing directly into waves is known to be challenging, therefore it is useful to demonstrate an ability within the routing algorithm to avoid this occuring.

Three different simulations were run with the three different failure criteria. Figure 4 shows the shortest path where the failure model allows for two failure criteria to be met, it can be seen that the shortest path travels directly through both areas. The single failure criteria route avoids area 1 but travels through area 2, as shown in Figure 5. The double failure criteria route, shown in Figure 6, shows that the failure model is able to avoid both areas. The route takes 29 days and 7 hours ignoring any failure criteria, with the time taken increasing by 25 hours to avoid two failure criteria and by another 25 hours in order to avoid any failure criteria.

The results illustrated in Figures 4 to 6 illustrate the ability for the routing algorithm to avoid areas where the failure model calculates a probability which is unacceptable. Each routing result differs by an amount significantly over 2.1 hours, the estimated accuracy of the original routing algorithm, indicating that the variation in result is due to the failure model.

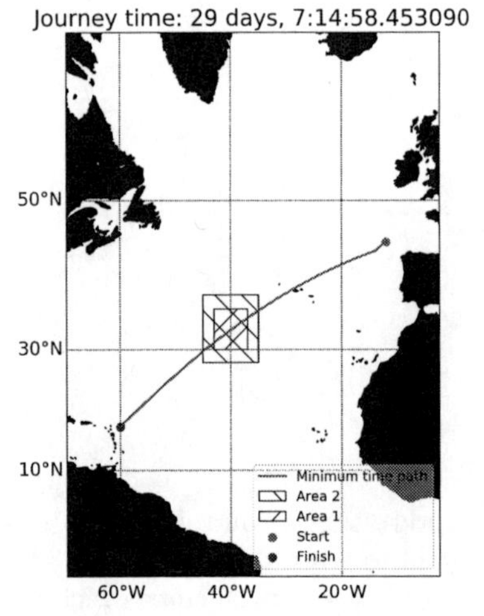

Figure 4: Shortest path avoiding any failures.

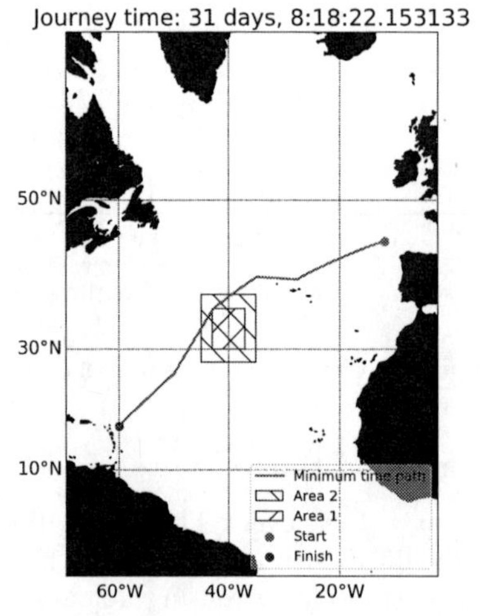

Figure 5: Shortest path avoiding two failures.

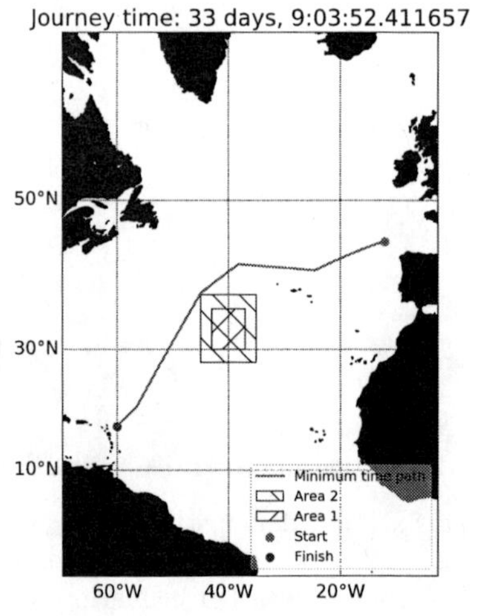

Figure 6: Shortest path avoiding one failure.

4 Conclusions

Autonomous sailing craft competing the MicroTransat competition have suffered some form of critical failure before they have been able to finish. To model this problem this paper has introduced a novel methodology by introducing a failure model within the sailing craft routing algorithm. In order to demonstrate the impact of the routing model the accuracy of the routing algorithm needed to be quantified.

Routing algorithms are generally based on a discretization of a continuous domain that results in an error. This error is reduced as the accuracy of the discretization process is increased but increases the computational cost of running the simulation. The error was calculated using the grid convergence index, a parameter borrowed from the discipline of Computational Fluid Dynamics and measures the difference between the simulated result and the actual result inferred based of a trend of previous results. The grid convergence index was calculated for a series of routing simulations which were solved for a range of different grid sizes. A balance between computational run time and accuracy was achieved through using an effective grid height of 36 km to discretize the domain. This corresponds to an error of ± 2.1 hours over a voyage lasting 703.25 hours.

A failure model has been implemented that is able to model two levels of failure, thereby demonstrating the flexibility of the Bayesian Belief network with regards to modelling different combinations of failure causes. The levels of failure modelled are a function of the mean wave height and direction. This has modelled the difficulty for sailing craft to maintain a course when either sailing in heavy seas or sailing into the wave direction.

To demonstrate the ability of the failure model to avoid areas which exceed a specified probability of failure, simulations varying the probability of failure were conducted with a control weather scenario. In the centre of the domain two patches of failure-inducing weather were placed. The failure model routing algorithm was able to avoid these failure inducing patches according to different acceptable levels of failure. To the authors knowledge this is the first calculation and application of discretization error in the interpretation of routing algorithm simulation results.

4.1 Future work

Integrating empirical or modelled autonomous sailing craft failure data into the failure model would allow realistic route modelling to take place. This will involve collaboration with designers and users who have been able to model or record specific failure modes and their causes and to incoporate this information into an improved failure model. As the failure model has demonstrated an ability to avoid areas of risk, one extentsion could be to input the locations of common fishing areas, as a common cause of ASC failure is being caught by fishing vessels (Microtransat Challenge, 2018).

The routing algorithm does not account for the time cost associated with tacking or gybing. It is noted that the time taken for an ASV to recover speed is small with respect to the overall time of voyaging, however as the accuracy of the routing has been quantified it is now possible to test whether a significant difference is achieved through including such a manouevre model.

Another avenue of research could be into the use of Genetic Algorithms or Monte Carlo Tree Search method for the purpose of accelerating the solution optimisation time.

4.1.1 Acknowledgements

Thanks must go to the University of Southampton Sailing Robot team for their support and encouragement.

The code developed to produce the simulations in this report has been developed under the MIT licence and can be found online at `https://github.com/TAJD/pyroute`.

References

Auboin, L., Blake, J. I. R. and Turnock, S. R. (2010). A fault tree based investigation of the reliability of ocean racing yachts incorporating human performance and canting keel impacts. In *2nd International Conference on Innovation in High Performance Sailing Yachts*.

Bellman, R. (1957). *Dynamic Programming*. Princeton University Press, Princeton, NJ, republishe edition.

Brito, M. and Griffiths, G. (2016). A Bayesian approach for predicting risk of autonomous underwater vehicle loss during their missions. *Reliability Engineering & System Safety*, 146:55–67.

Celik, I. B., Ghia, U., Roache, P. J., Freitas, C. J., Coleman, H., and Raad, P. E. (2008). Procedure for Estimation and Reporting of Uncertainty Due to Discretization in CFD Applications. *Journal of Fluids Engineering*, 130(7):078001.

Dalang, R. C., Dumas, F., Sardy, S., Morgenthaler, S., and Vila, J. (2014). Stochastic optimization of sailing trajectories in an upwind regatta. *Journal of the Operational Research Society*, pages 1–15.

Dhomè, U. (2017). Further development and performance evaluation of the autonomous sailing boat Maribot Vane. Technical report, KTH, Royal Institute of Technology 2017.

Ferretti, R. and Festa, A. (2018). A Hybrid control approach to the route planning problem for sailing boats. *http://arxiv.org/abs/1707.08103*, pages 1–27.

Friis-Hansen, A. (2000). *Bayesian networks as a decision support tool in marine applications.* Phd, Technical University of Denmark.

Helvacioglu, S. and Ozen, E. (2014). Fuzzy based failure modes and effect analysis for yacht system design.

Ladany, S. P. and Levi, O. (2017). Search for optimal sailing policy. *European Journal of Operational Research*, 260(1):222–231.

Microtransat Challenge (2018). Microtransat Challenge.

Philpott, A. and Mason, A. (2001). Optimising yacht routes under uncertainty. *In Proc. of the 15th Chesapeake Sailing Yacht Symposium, Annapolis, MD*, pages 1–8.

Philpott, A. B., Henderson, S. G., and Teirney, D. (2004). A Simulation Model for Predicting Yacht Match Race Outcomes. *Operations Research*, 52(1):1–16.

Roache, P. J. (1997). Quantification of Uncertainty in Computational Fluid Dynamics. *Annual Review of Fluid Mechanics*, 29(1):123–160.

Schlaefer, A. and Blaurock, O. (2011). Route Planning for a Micro-transat Voyage. In Schlaefer, A. and Blaurock, O., editors, *Robotic Sailing: Proceedings of the 4th International Robotic Sailing Conference.* Springer.

Spenkuch, T. B. (2014). *A Bayesian Belief Network Approach for Modelling Tactical Decision-Making in a Multiple Yacht Race Simulator.* PhD thesis, University of Southampton.

Stelzer, R. and Pröll, T. (2008). Autonomous sailboat navigation for short course racing. *Robotics and Autonomous Systems*, 56(7):604–614.

Tagliaferri, F., Philpott, A. B., Viola, I. M., and Flay, R. G. J. (2014). On risk attitude and optimal yacht racing tactics. *Ocean Engineering*, 90:149–154.

Tagliaferri, F. and Viola, I. M. (2017). A real-time strategy-decision program for sailing yacht races. *Ocean Engineering*, 134(February):129–139.

Study of Long-term Route Planning for Autonomous Sailboat

Mingshu Du
Shanghai Jiao Tong University
Shanghai, China
dms1415@163.com

Mengqi Kang
Seastel Marine System (Shanghai) Co. Ltd.
Shanghai, China
kangmengqi@seastel.com

Chunxiao Hou
Seastel Marine System (Shanghai) Co. Ltd.
Shanghai, China
houchunxiao@seastel.com

Jinsong Xu
Shanghai Jiao Tong University
Shanghai, China
jinsong@sjtu.edu.cn

Abstract

The multi-dimensional dynamic programming method is applied to the long-term route planning of an autonomous sailboat from Shanghai to Qingdao. The sailed voyage length out of the current position is adopted to be the third-dimensional state variable. The short-term route planning between the neighboring waypoints is defined as the control variable. A group of optimal routes with minimum voyage time can be obtained corresponding to different voyage length. The result shows that by adding the state variable dimension, more optional routes are kept in the final results, which is valuable for better decision support.

1 Introduction

Marine environment monitoring has great significance in sustainable development of marine economy. With improvement of telecommunication and sensor technology, marine monitoring methods have gradually changed from exploration mode which relies on a large number of human resources to observation mode with longer monitoring time and wider coverage. The sailboat relying on natural wind to keep autonomous navigation is suitable for carrying various sensors to accomplish different monitoring tasks (Cruz, 2008) (Rynne, 2009).
In the case that the destination and the wind field in the sailing area are known, the conventional motor boat must avoid strong wind area as much as possible to ensure safety and to reduce fuel consumption. However, route planning for sailboats must fully consider favorable wind conditions. Therefore, long-term route planning of the sailboat is more complex and important than that of conventional motor boats. The most popular long-term route planning method for sailboats is based on A* algorithm. Ulm University in Germany developed a route planning program based on A* algorithm and wind field forecast information (Langbein, 2011). Shanghai Jiao Tong University adopted A* algorithm and local optimal method respectively to plan the long-term route (Kang, 2016). The planning objectives of those researches are minimum voyage time.

In: S. M. Schillai, N. Townsend (eds.): Proceedings of the International Robotic Sailing Conference 2018, Southampton, United Kingdom, 31-08-2018

Since only a few variables can be considered in the planning process of A* algorithm, it is difficult to handle complex environmental factors and constraints. United States Naval Academy selected Northern Route and Southern Route as two feasible plans for "Spirit of Annapolis" transatlantic navigation based on the pilot charts, and then compared the strengths and weaknesses of the two routes according to hurricanes, currents, sea ice, wind fields, and seaports (Gibbons-Neff, 2011). The factors considered in that planning process are far more than those of the conventional A* algorithm. If planning algorithms are used to assist decision-making in route planning, more diverse planning results should be provided. Strathclyde University in the United Kingdom adopted three-dimensional dynamic programming (3DDP) in the route planning of motor ships. By defining position and sailing time as state variables, heading and sailing speed as decision variables, a group of minimum fuel consumption paths with different sailing time was planned for further consideration (Shao, 2013).

In this paper, 3DDP was applied for long-term route planning of an unmanned sailboat from Shanghai to Qingdao. The current position and the sailed voyage length are defined as state variables. The short-distance route planning between the neighbor waypoints is defined as the control variable. For different total lengths, a group of routes with minimum voyage time is obtained as decision assistance. The final long-term solution can be chosen from those planned routes by combining more factors.

2 Sailboat Model and Sailing Sea Area

2.1 Sailboat Model and Polar Diagram of sailboat Speed

The total length of the sailboat model is 1.5m. Based on the existing hull design scheme (Wang, 2015), two rigid wing sails are equipped and the airfoil profile is NACA0012. The overall appearance of the sailboat is shown in Figure 1. Static VPP algorithm is used to obtain the polar diagram of sailboat speed (Oossanen, 1993). The curves shown in Figure 2 respectively represent the maximum speed of different heading angles when wind direction is 0 degree and wind speed is V_{wind}. The VPP results are not only rapidity indicator but the basis for estimating sailing time in long-term route planning.

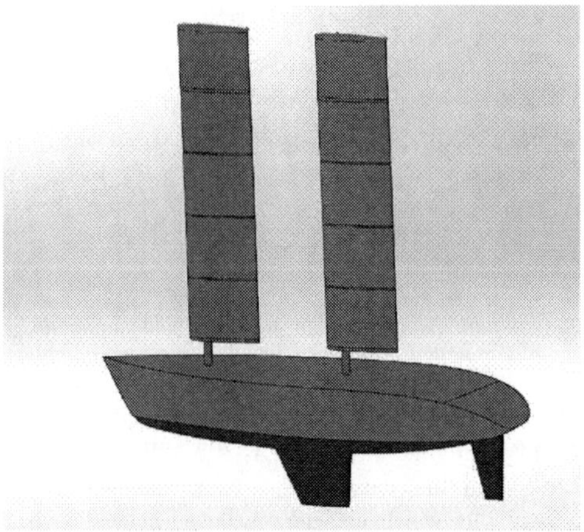

Figure 1: Sailboat model

2.2 Grid Design

In long-term route planning for sailboat, the sailing area is discretized into a grid system to specify the spatial layout of stages and states. As described in Figure 3, the great circle route which represents the shortest course from Shanghai to Qingdao on the surface of the earth is divided into $(N-1)$ stages equally, and M points are created perpendicularly away from the great circle with a unit spacing of X. Thus, the grid is described as (i,k). For example, the departure is $((M+1)/2,1)$, and the destination is $((M+1)/2,N)$. In this paper, the parameters are N=31 and M=31.

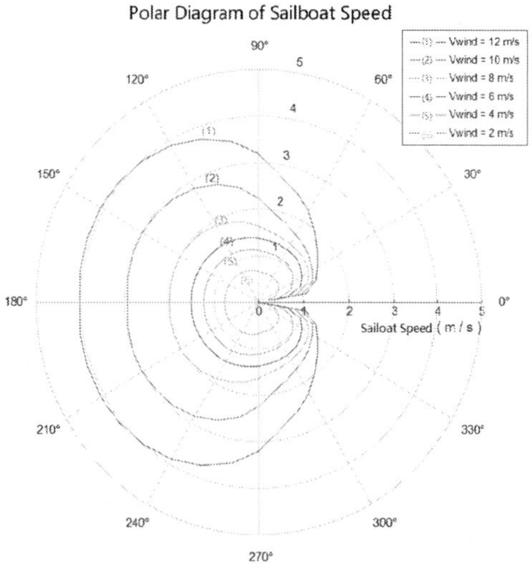

Figure 2: Polar diagram of sailboat speed

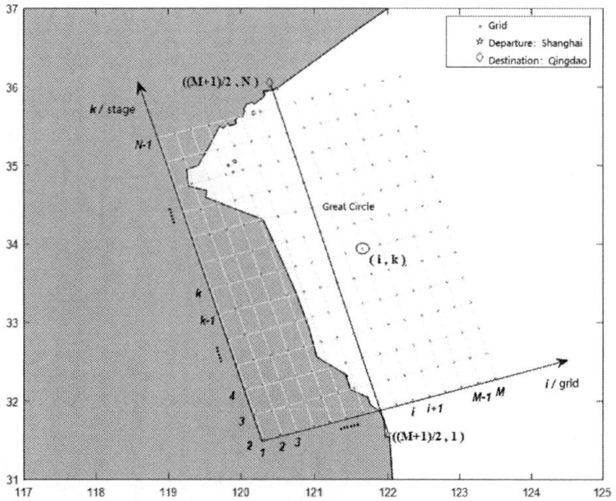

Figure 3: Projections of the space grid system

3 Dynamic Programming for sailboat route planning

Dynamic programming is an effective method to solve multi-stage decision problems. By establishing the recursive relationship between two neighboring stages, the optimal decision at each stage can be obtained. The optimal decision at each stage establishes the optimal decision sequence for the entire process (Teng, 2011). There are two kinds of solutions in dynamic programming. Backward dynamic programming is recursive from the destination to the starting stage, and forward dynamic programming is recursive from the starting point to the destination. The recursive relationship of forward dynamic programming is described as (Shao, 2013):

$$J^*(\overrightarrow{X}(k),k) = \min_{\overrightarrow{U}(q-1),q=2,\cdots,k} \{\sum_{q=2}^{k} \alpha_q(\overrightarrow{X}(q),\overrightarrow{U}(q-1),q)\}$$
$$= \min_{\overrightarrow{U}(k-1)} \{\alpha_k(\overrightarrow{X}(k),\overrightarrow{U}(k-1),k) + J^*(\overrightarrow{X}(k-1),k-1)\}$$
$$J^*(\overrightarrow{X}(1),1) = 0$$
$$k = 2,3,\cdots,N$$

where k is the stage variable. $\overrightarrow{X}(q)$ is the state variable of stage k. $\overrightarrow{U}(q-1)$ is the control variable which can make the sailboat transfer from state $\overrightarrow{X}(q-1)$ at the stage $(q-1)$ to state $\overrightarrow{X}(q)$ at stage q. $\alpha_q(\overrightarrow{X}(q),\overrightarrow{U}(q-1),q)$ is the cost function from stage $(q-1)$ to q. $\sum_{q=2}^{k} \alpha_q(\overrightarrow{X}(q),\overrightarrow{U}(q-1),q)$ is the objective function at the current stage k. The control sequence $\overrightarrow{U}(q-1)$ corresponding to the minimum objective function $J^*(\overrightarrow{X}(k),k)$ is the best decision sequence.
The recursive relationship between $J^*(\overrightarrow{X}(k),k)$ and $J^*(\overrightarrow{X}(k-1),k-1)$ in the equation above reflects the multi-stage thinking of dynamic programming. From the minimum of $J^*(\overrightarrow{X}(k-1),k-1)$ at stage $(k-1)$ and all cost function $\alpha_k(\overrightarrow{X}(k),\overrightarrow{U}(k-1),k)$, the optimal control variable $\overrightarrow{U}(k-1)$ and the minimum objective function $J^*(\overrightarrow{X}(k),k)$ at stage k can be obtained. One dimensional or multi-dimensional programming can be used depending on the dimension of the state variable $\overrightarrow{X}(q)$.
In this research, the stage variable is $k = 1,2,3,\cdots,31$. k=1 is in Shanghai and k=31 is in the destination Qingdao. In the state variable $X(i,j,k)$, (i,k) represents the grid point shown in Figure3, and j is a state variable of the route length $L_{i,k}$ from the departure to the current waypoint (i,k). The route length $L_{i,k}$ is equally devided into 30 groups labelled with the state variable j by using the following rounding equation:

$$j = \left\lfloor \frac{L_{i,k} - L_{k,min}}{(L_{k,max} - L_{k,min})/30} + 1 \right\rfloor$$

where $L_{k,min}$ is the minimum route length of the great circle route from the departure to the current stage k, and $L_{k,max}$ is the double value of $L_{k,min}$. After division of the route length by stete variable j, all the routing scheme from the departure to the current waypoint are divided into 30 groups and from each group one optimal route plan can be obtained. Obviously, the additional dimension of the state variable j provides more alternatives for decision support. The scheme is illustrated in Figure 4.

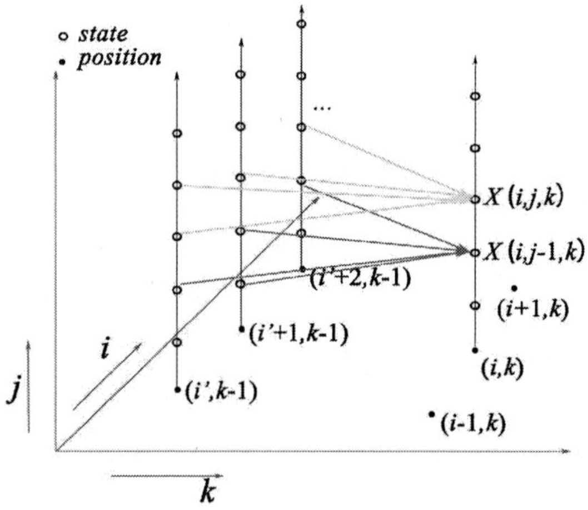

Figure 4: Illustration of state variable $X(i,j,k)$

The control variable $U(i',j',k-1)$ which makes the sailboat transfer from the state variable $X(i',j',k-1)$ at stage $(k-1)$ to the state variable $X(i,j,k)$ at stage k is actually the short-term routing scheme from the waypoint

$(i', k-1)$ to (i, k). In this paper, the straight path scheme with the maximum speed is used for downwind and beamwind sailing, and the tacking scheme with the maximum speed is used for sailing against wind.
The cost function $t(X(i,j,k), U(i',j',k-1), k)$ is the sailing time from waypoint $(i', k-1)$ to (i, k) under the control variable $U(i',j',k-1)$. The maximum speed when sailing downwind or beamwind is obtained according to polar diagram of sailboat speed and wind field. When sailing against wind the maximum speed is defined as $C_{tack}\Delta u_{\pi/4}$ where C_{tack} is the cost coefficient of tacking and $u_{\pi/4}$ is the maximum speed when wind direction is $\pi/4$. Thus, the dynamic recurrence relation of the long-term route planning of the sailboat is expressed as below, and the detailed process of dynamic planning is shown in Figure 5.

$$J_p^*(X(i,p,k),k) = \min_{U(i',j',k-1),j=p} \{t(X(i,j,k), \vec{U}(i',j',k-1),k) + J_j^*(\vec{X}(i',j',k-1),k-1)\}$$

$$p = 1, 2, \cdots, 30$$

$$i = 1, 2, \cdots, 31$$

$$J_1^*(X(16,1,1),1) = 0$$

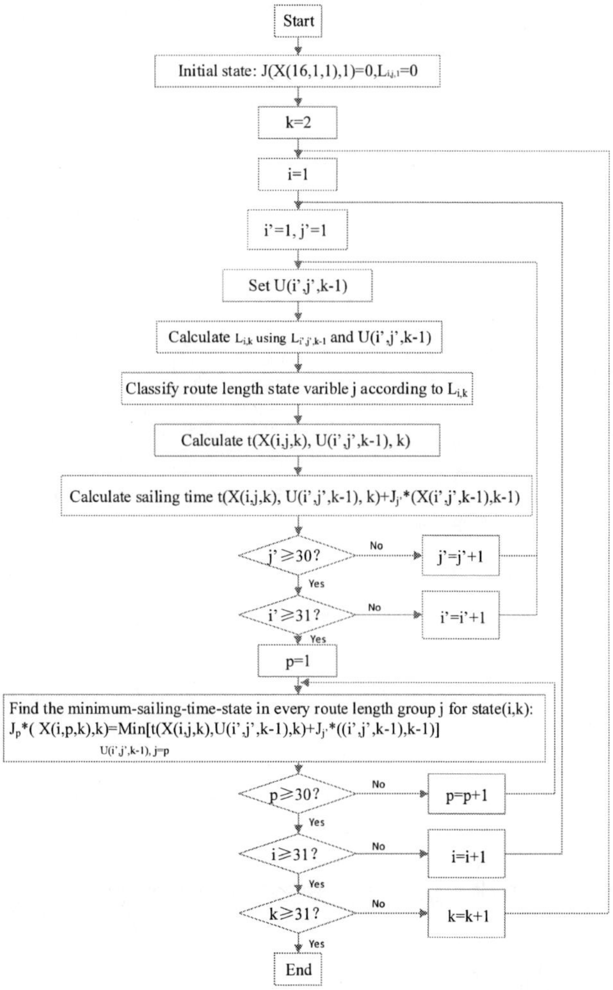

Figure 5: Flow of long-term route planning for sailboat

4 Long-term Route Planning Results

4.1 Steady Uniform Wind Field

3DDP was adopted in the route planning from Shanghai to Qingdao. The wind speed was 6m/s steadily and the wind direction was north, south, west and east respectively. As shown in Figure 6, a group of routes with the shortest time was obtained in different route length. In the cases of south, east and west wind, the total length of the route is positively correlated with the minimum voyage time. The route with the shortest length and sailing time is almost a straight line between the starting point and the destination. The minimum voyage time for those three cases is about 100 hours. With the total length increasing, the route with the minimum voyage time gradually moves away from the coastline. Obviously, by adding the total length of the route as a planning variable, the planning result provides more selections and plays a better role in decision support.

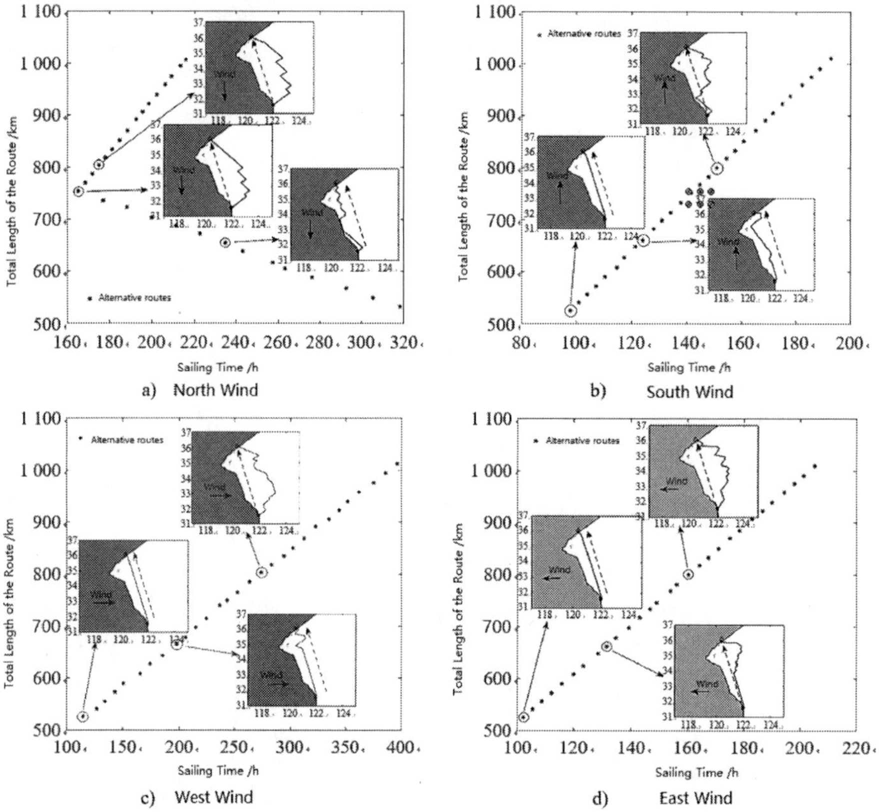

Figure 6: Route planning results in steady uniform wind field

In the case of north wind, the minimum voyage time is about 160 hours with a total route length of 750km, which is significantly longer than that of other three cases. The total length of the route is positively correlated with the minimum voyage time when the route length is over 750km. However, when the route length is less than 750km, the minimum voyage time increases significantly with the decreasing of route length. The short route less than 750km restricts the tacking operation to a smaller upwind angle which causes slower speed and longer sailing time.

4.2 Steady non-uniform Wind Field

As shown in Figure 7, the instantaneous wind field on May 20th, 2017 was adopted according to National Centers for Environmental Prediction. The maximum wind speed is about 10m/s. A group of alternative routes with minimum voyage time in different voyage length from Shanghai to Qingdao was obtained by using 3DDP method.

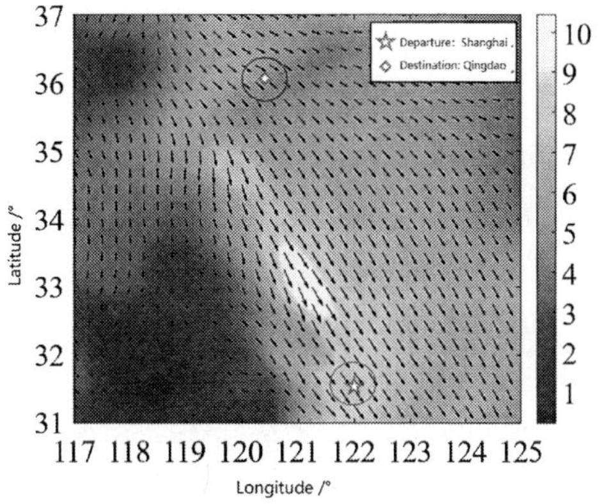

Figure 7: Wind direction and Speed of steady non-uniform wind field

The simulation result in Figure 8 is similar to that of steady north wind. The minimum voyage time is 150h corresponding to a total voyage length of 750km. But the route is close to the coastline and is risky of collision. Thus, the final sailing route could be selected from the other alternatives.

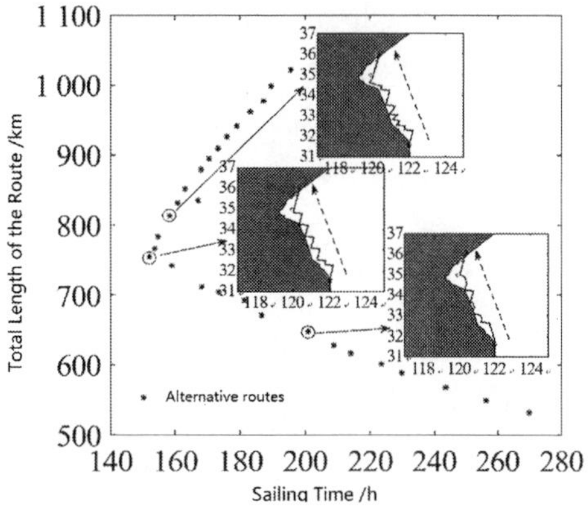

Figure 8: Route planning results in steady non-uniform wind field

5 Conclusions

In this research, 3DDP is adopted to long-term route planning for an autonomous sailboat. Apart from the current position of the sailboat, the sailed voyage length is defined as the third state variable. Without that variable, only one optimal path is obtained from all the path schemes. However, in 3DDP method, all path schemes from the departure to the current point are divided into up to 30 groups according to the total length of the route. Each group can generate a planning result with minimum voyage time. 3DDP provides more alternatives for decision support.

When the sailboat sails downwind or beamwind, the total length of the route is positively correlated with the minimum voyage time. The route with the shortest length and sailing time is almost a straight path between the departure and the destination.

When the sailboat sails against wind, if the total length of the route is less than a certain critical value, the minimum sailing time will increase significantly with the total length of the route decreasing restricted by tacking operation.

References

Cruz, N.A. (2008). Autonomous Sailboats: an Emerging Technology for Ocean Sampling and Surveillance. Oceans. IEEE, 2008:1-6.

Gibbons-Neff, P. (2011). Route Planning for a Micro-transat Voyage. Robotic Sailing, 2011:183-194.

Kang, M. (2016). Study on Route Planning for Autonomous Sailboat. Shanghai: Shanghai Jiao Tong University, 2016

Langbein, J. (2011). A Rule-Based Approach to Long-Term Routing for Autonomous Sailboats. Robotic Sailing. Springer Berlin Heidelberg, 2011:195-204.

Oossanen, P. (1993). Predicting the Speed of Sailing Yachts. Discussion. Author's Closure. Transactions-society of Naval Architects and Marine Engineers, 1993, 101: 337-397.

Rynne, P.F. (2009). Unmanned Autonomous Sailing: Current Status and Future Role in Sustained Ocean Observations. Marine Technology Society Journal, 2009, 43(1): 21-30.C

SHAO, W. Development of an Intelligent Tool for Energy Efficient and Low Environment Impact Shipping. University of Strathclyde, Glasgow, 2013.

Teng, Y. (2011). Principle and Appliance of Dynamic Route Planning. Chengdu: Southwest Jiao Tong University Press, 2011: 1-16.

Wang, Q. (2015). Autonomous Sailboat Track Following Control. Proceedings of the 8th International Robotic Sailing Conference, 2015: 125-136.

Tight Slalom Control for Sailboat Robots

Mael Le Gallic
ENSTA Bretagne
mael.le_gallic@ensta-bretagne.org

Joris Tillet
ENSTA Bretagne
joris.tillet@ensta-bretagne.org

Fabrice Le Bars
Lab-STICC
ENSTA Bretagne
fabrice.le_bars@ensta-bretagne.fr

Luc Jaulin
Lab-STICC
ENSTA Bretagne
lucjaulin@gmail.com

Abstract

Existing controllers for sailboat robots are usually developed for speed performances and for long straight lines. In this context, the accuracy is not the main concern. In this paper, we consider the tight slalom problem which requires accuracy. We propose a feedback-linearization based method combined with a vector field approach to control the sailboat. Some simulations show that the robot is able to perform the slalom without missing any gate.

1 Introduction

We consider a mobile robot described by the state equations (Jaulin, 2015a)

$$\begin{cases} \dot{\mathbf{x}} &= \mathbf{f}(\mathbf{x}, \mathbf{u}) \\ \mathbf{p} &= \mathbf{g}(\mathbf{x}) \end{cases} \tag{1}$$

with an input vector $\mathbf{u} = (u_1, \ldots, u_m)$, a state vector $\mathbf{x} = (x_1, \ldots, x_n)$ and a pose vector $\mathbf{p} = (p_1, \ldots, p_{m+1})$ with $n \geq m + 1$. The goal of this paper is to show we can follow a chosen vector field in the $\mathbf{p}$ space (Khatib, 1986)(Pêtres et al., 2011)(Schmitt et al., 2016), using a feedback-linearization based method. It means that we can control $m + 1$ state variables and not only m of them, as given by the theory (Isidori, 1995). This is due to the fact that we perform a path following instead of a trajectory tracking where the time is involved. In practice, the vector $\mathbf{p}$ corresponds to the position of the center of the robot and may be of dimension 2 (if $m = 1$) or 3 (if $m = 2$). This is consistent with the fact that we need one actuator to control the direction of a 2D vehicle such as a car or a boat and two actuators for a 3D vehicle such as a plane.

The approach we propose is to find a controller so that the vector $\dot{\mathbf{p}}$ be collinear (instead of equal) to the required field. This is illustrated in this paper in the case where the mobile robot is a sailboat (Miller et al., 2012)(Cruz and Alves, 2008)(Holger et al., 2009). The input $\mathbf{u}$ is scalar (*i.e.*, $m = 1$) and corresponds to the rudder. Moreover, we will show that this approach is particularly adapted to sailboats where the speed is hardly controllable (Neumann and Schlaefer, 2012)(Stelzer et al., 2007).

In: S. M. Schillai, N. Townsend (eds.): Proceedings of the International Robotic Sailing Conference 2018, Southampton, United Kingdom, 31-08-2018

2 Method

In order to facilitate the understanding of our approach, we will deal with a Dubins car, which is much simpler than a sailboat. The extension to other type of mobile robots is straightforward.

2.1 Line following for a Dubins car

To introduce our approach, we consider a robot (here a Dubins car) moving on a plane and described by the following state equations:

$$\left\{\begin{array}{lll}\dot{x}_1 & = & \cos x_3 \\ \dot{x}_2 & = & \sin x_3 \\ \dot{x}_3 & = & u\end{array}\right. \tag{2}$$

where x_3 is the heading of the robot and $\mathbf{p} = (x_1, x_2)$ are the coordinates of its center. The state vector is given by $\mathbf{x} = (x_1, x_2, x_3)$.

Let us choose as the control output the variable

$$y = x_3 + \operatorname{atan}(x_2). \tag{3}$$

and let us find a classical feedback linearization based controller (Jaulin, 2015b) such that the output y (which can be interpreted as an error) converges to 0. In such a case, we will have

$x_3 + \operatorname{atan}(x_2) = 0$

and the robot will perform a line following. Differentiating (3) we have

$$\dot{y} = \dot{x}_3 + \frac{\dot{x}_2}{1 + x_2^2} = u + \frac{\sin x_3}{1 + x_2^2}. \tag{4}$$

Since u occurs in (4), the relative degree of the system is 1. We may thus choose a first order equation for the error y, such as

$$\dot{y} + y = 0, \tag{5}$$

We then choose u to have this error equation satisfied. From (4) and (5), we get:

$$\begin{array}{lll} u & = & -y - \frac{\sin x_3}{1+x_2^2} \\ & = & -x_3 - \operatorname{atan}(x_2) - \frac{\sin x_3}{1+x_2^2} \end{array} \tag{6}$$

Note that we do not have any singularity. As illustrated by the simulation depicted on Figure 1, the associated vector field makes the car attracted by the line $x_2 = 0$.

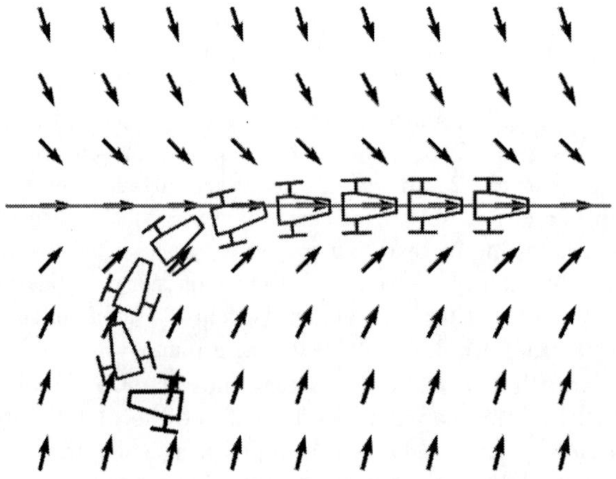

Figure 1: Precise line following

Remark. For more robustness with respect to small uncertainties, a sliding mode effect could be added. It suffices to take for required error

$$y = x_3 + \operatorname{atan}\left(x_2 + \alpha \cdot \operatorname{sign}\left(x_2\right)\right),$$

where α is a small positive coefficient, *e.g.*, $\alpha = 0.1$. In such a case, the robot will go to the line in a finite time (and not asymptotically, as previously). Moreover, it will remains exactly on the line even if some small uncertainties occur.

2.2 Generalization

We want our robot to follow the field $\psi(\mathbf{p})$, more precisely, we want that $\psi(\mathbf{p})$ and $\dot{\mathbf{p}}$ point toward the same direction. This condition can be translated into the form $\varphi(\psi(\mathbf{p}), \dot{\mathbf{p}}) = 0$, where φ is a *collinearity function* which satisfies

$$\varphi(\mathbf{r}, \mathbf{s}) = \mathbf{0} \Leftrightarrow \exists\lambda > 0,\ \lambda\mathbf{r} = \mathbf{s}. \tag{7}$$

Typically, this function corresponds to one angle (the heading) if $m = 1$ and two angles (heading, elevation) for $m = 2$. Note that the function φ cannot be expressed with a determinant since $\mathbf{r}, \mathbf{s}$ should not point toward opposite directions. We define the output

$$\mathbf{y} = \varphi(\psi(\mathbf{p}), \dot{\mathbf{p}}) = \varphi\left(\psi(\mathbf{g}(\mathbf{x})), \frac{\partial \mathbf{g}}{\partial \mathbf{x}}(\mathbf{x}) \cdot \mathbf{f}(\mathbf{x}, \mathbf{u})\right). \tag{8}$$

Since $\mathbf{y} \in \mathbb{R}^m$, we can apply a feedback linearization method and we get $\mathbf{y} \to \mathbf{0}$. This means that the robot will follows the required field. Note that we have no control on the speed, which is not our main concern in this paper.

2D case. Consider for instance the case where $m = 1$. We have

$$\psi(\mathbf{p}) = \begin{pmatrix} \psi_1(\mathbf{p}) \\ \psi_2(\mathbf{p}) \end{pmatrix}. \tag{9}$$

We take as an output y, the angle between the actual heading vector $\dot{\mathbf{p}} = \frac{\partial \mathbf{g}}{\partial \mathbf{x}}(\mathbf{x}) \cdot \mathbf{f}(\mathbf{x}, \mathbf{u})$ and the desired heading vector given by $\psi(\mathbf{p})$. Denote by $\theta(\mathbf{x})$ the argument of the vector $\dot{\mathbf{p}}$. We have

$$\begin{array}{rcl} y \overset{(8)}{=} & = & \text{angle}\left(\psi(\mathbf{g}(\mathbf{x})), \frac{\partial \mathbf{g}}{\partial \mathbf{x}}(\mathbf{x}) \cdot \mathbf{f}(\mathbf{x}, \mathbf{u})\right) \\ & = & \text{angle}\left(\psi(\mathbf{g}(\mathbf{x})), \begin{pmatrix} \cos\theta \\ \sin\theta \end{pmatrix}\right) \\ & = & \text{sawtooth}(\theta - \text{atan2}(\underbrace{\psi_2(\mathbf{g}(\mathbf{x}))}_{b}, \underbrace{\psi_1(\mathbf{g}(\mathbf{x}))}_{a})) \end{array} \tag{10}$$

The *sawtooth* function is given by:

$$\text{sawtooth}(\widetilde{\theta}) = 2\text{atan}\left(\tan\tfrac{\widetilde{\theta}}{2}\right) = \text{mod}(\widetilde{\theta} + \pi, 2\pi) - \pi \tag{11}$$

As illustrated in Figure 2, the function corresponds to an error in heading. The interest in taking an error $\widetilde{\theta}$ filtered by the *sawtooth* function is to avoid the problem of the $2k\pi$ modulus: we would like a $2k\pi$ to be considered non-zero.

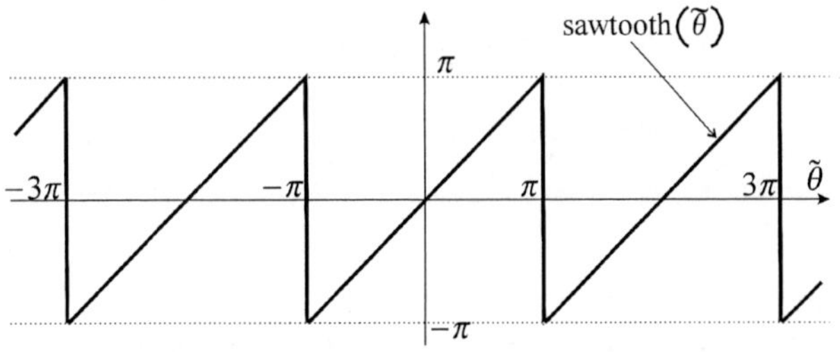

Figure 2: *Sawtooth* function used to avoid the jumps in the heading control

We have

$$\begin{array}{rcl} \dot{y} & = & \dot{\theta} - (\underbrace{-\frac{b}{a^2+b^2}}_{\frac{\partial \text{atan2}(b,a)}{\partial a}} \cdot \dot{a} + \underbrace{\frac{a}{a^2+b^2}}_{\frac{\partial \text{atan2}(b,a)}{\partial b}} \cdot \dot{b}) \\ & = & u + \frac{b \cdot \dot{a} - a \cdot \dot{b}}{a^2+b^2}, \end{array} \tag{12}$$

if we assume that the input u corresponds to the desired angular velocity. We propose a feedback linearization based control based on the required equation $\dot{y} = -y$. We have

$$\begin{array}{lll} u & \overset{(12)}{=} & \dot{y} - \frac{\left(b\cdot\dot{a}-a\cdot\dot{b}\right)}{a^2+b^2} \\ & = & -y - \frac{\left(b\cdot\dot{a}-a\cdot\dot{b}\right)}{a^2+b^2} \qquad\qquad \text{(since } \dot{y} = -y\text{)} \\ & \overset{(10)}{=} & -\left(\text{sawtooth}(\theta - \text{atan2}(b,a)\right) - \frac{\left(b\cdot\dot{a}-a\cdot\dot{b}\right)}{a^2+b^2}. \end{array} \tag{13}$$

We thus have the guarantee that after some time, the error angle y is 0 and that we follow exactly the vector field.

2.3 Dubins car following the Van der Pol cycle

We would like our Dubins car to follow a path corresponding to the limit cycle of the Van der Pol equation:

$$\psi(\mathbf{p}) = \begin{pmatrix} p_2 \\ -\left(0.01\ p_1^2 - 1\right) p_2 - p_1 \end{pmatrix}. \tag{14}$$

Take $\mathbf{g}(\mathbf{x}) = (x_1, x_2)^{\mathrm{T}}$ which means that we want to build the paths in the (x_1, x_2)-space. We have

$$\psi(\mathbf{g}(\mathbf{x})) = \begin{pmatrix} x_2 \\ -\left(0.01\ x_1^2 - 1\right) x_2 - x_1 \end{pmatrix} \tag{15}$$

and

$$\frac{\partial \mathbf{g}}{\partial \mathbf{x}}(\mathbf{x}) \cdot \mathbf{f}(\mathbf{x}, \mathbf{u}) = \begin{pmatrix} 1 & 0 & 0 \\ 0 & 1 & 0 \end{pmatrix} \cdot \begin{pmatrix} \cos x_3 \\ \sin x_3 \\ u \end{pmatrix} \tag{16}$$

Thus

$$\begin{array}{lll} a & = & x_2 \\ b & = & -(0.01\ x_1^2 - 1)x_2 - x_1 \\ \theta & = & x_3 \end{array} \tag{17}$$

and

$$\begin{array}{lll} \dot{a} & = & \sin x_3 \\ \dot{b} & = & -\left(0.01 \cdot 2x_1\dot{x}_1\right) x_2 - \left(0.01\ x_1^2 - 1\right) \dot{x}_2 - \dot{x}_1 \\ & = & -0.02 \cdot x_1 x_2 \cos x_3 - \left(0.01\ x_1^2 - 1\right) \sin x_3 - \cos x_3 \end{array} \tag{18}$$

From (13), we get that final controller is

$$\begin{array}{lll} u & = & -\text{sawtooth}\left(x_3 - \text{atan2}\left(-\left(\frac{x_1^2}{100} - 1\right) x_2 - x_1, x_2\right)\right) \\ & & + \frac{\left(\left(\frac{x_1^2}{100}-1\right)x_2+x_1\right)\cdot\sin x_3 + x_2\cdot\left(\frac{x_1 x_2 \cos x_3}{50} + \left(\frac{x_1^2}{100}-1\right)\sin x_3 + \cos x_3\right)}{x_2^2 + \left(\left(\frac{x_1^2}{100}-1\right)x_2+x_1\right)^2} \end{array} \tag{19}$$

The behavior of the control law is illustrated by Figure 3. The car is very close to the true limit cycle, which is not the case if we consider a classical linear controller. Indeed, the controller anticipates the fact that the required trajectory have to take into account the curvature of the vector field.

3 Application to the slalom problem

We consider the following model which corresponds to a simplified version of the sailboat model given in (Jaulin and Le Bars, 2013). The state equations are

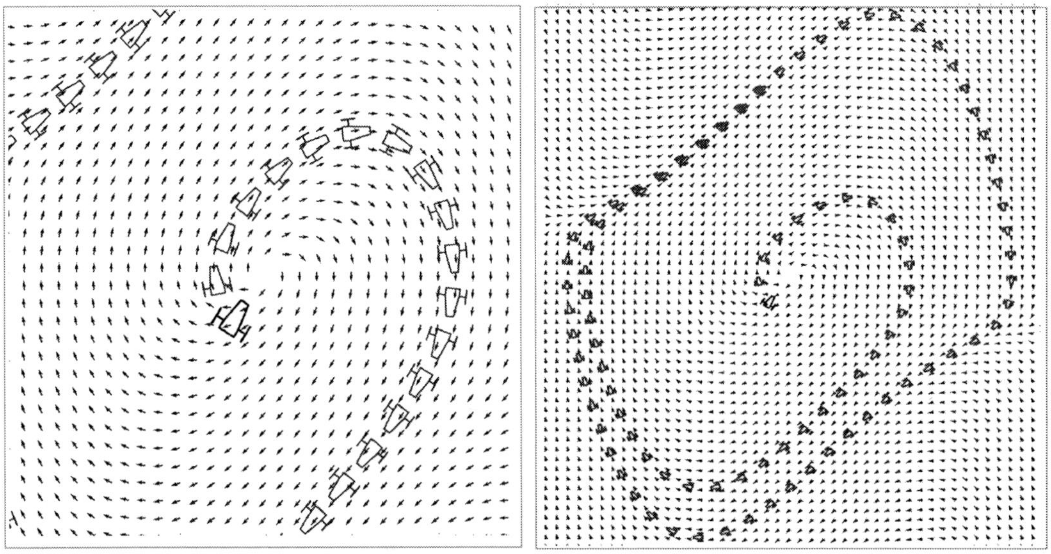

Figure 3: Dubins describing accurately the Van der Pol cycle

$$\left\{\begin{array}{lll}\dot{x}_1 & = & v\cos\theta \\ \dot{x}_2 & = & v\sin\theta \\ \dot{\theta} & = & -\rho_2 v\sin 2u_1 \\ \dot{v} & = & \rho_3\left\|\mathbf{w}_{\text{ap}}\right\|\sin\left(\delta_s-\psi_{\text{ap}}\right)\sin\delta_s-\rho_1 v^2 \\ \sigma & = & \cos\psi_{\text{ap}}+\cos u_2 \\ \delta_s & = & \left\{\begin{array}{cc}\pi+\psi_{\text{ap}} & \text{if } \sigma\leq 0 \\ -\text{sign}\left(\sin\psi_{\text{ap}}\right)\cdot u_2 & \text{otherwise}\end{array}\right. \\ \mathbf{w}_{\text{ap}} & = & \left(\begin{array}{c}-a\ \sin(\theta)-v \\ -a\ \cos(\theta)\end{array}\right) \\ \psi_{\text{ap}} & = & \text{angle } \mathbf{w}_{\text{ap}}\end{array}\right. \tag{20}$$

where $\rho_1=0.003,\rho_2=0.2,\rho_3=3$. In this equation u_1,u_2 correspond to the tuning of the rudder and the sail, respectively. We would like our robot to follow a path which makes a tight slalom through doors that have to be passed. We assume that we have a Cartesian equation for our path. For instance, we consider that the path is described by

$$e(\mathbf{p})=10\sin\left(\frac{p_1}{10}\right)-p_2=0 \tag{21}$$

where $e(\mathbf{p})$ corresponds to an error. This path corresponds to a path that should be possible for a normal sailboat robot for crosswind conditions. We take a vector field which corresponds to a pole placement strategy. For instance, we want the error satisfies $\dot{e}=-0.1\,e$, so that it will converge to zero in about 10 sec. Thus

$$\underbrace{\cos\left(\frac{p_1}{10}\right)\dot{p}_1-\dot{p}_2}_{\dot{e}(\mathbf{p})}=-\frac{1}{10}\underbrace{\left(10\sin\left(\frac{p_1}{10}\right)-p_2\right)}_{e(\mathbf{p})} \tag{22}$$

We take $\dot{p}_1=1$, to go to the right. As a consequence, we get the following field:

$$\boldsymbol{\psi}(\mathbf{p})=\left(\begin{array}{c}\dot{p}_1 \\ \dot{p}_2\end{array}\right)=\left(\begin{array}{c}1 \\ \cos\left(\frac{p_1}{10}\right)+\frac{1}{10}\left(10\sin\left(\frac{p_1}{10}\right)-p_2\right)\end{array}\right) \tag{23}$$

which is attracted by the curve $p_2=10\sin\left(\frac{p_1}{10}\right)$.

We have

$$\boldsymbol{\psi}(\mathbf{g}(\mathbf{x}))=\left(\begin{array}{c}1 \\ \cos\left(\frac{x_1}{10}\right)+\frac{1}{10}\left(10\sin\left(\frac{x_1}{10}\right)-x_2\right)\end{array}\right) \tag{24}$$

and

$$\frac{\partial \mathbf{g}}{\partial \mathbf{x}}(\mathbf{x}) \cdot \mathbf{f}(\mathbf{x}, \mathbf{u}) = \begin{pmatrix} \cos x_3 \\ \sin x_3 \end{pmatrix}. \quad (25)$$

Thus

$$\begin{array}{rcl} a & = & 1 \\ b & = & \cos\left(\frac{x_1}{10}\right) + \sin\left(\frac{x_1}{10}\right) - \frac{1}{10}x_2 \end{array} \quad (26)$$

and

$$\begin{array}{rcl} \dot{a} & = & 0 \\ \dot{b} & = & -\dot{x}_1 \frac{1}{10} \sin\left(\frac{x_1}{10}\right) + \dot{x}_1 \frac{1}{10} \cos\left(\frac{x_1}{10}\right) - \frac{1}{10}\dot{x}_2 \\ & = \frac{1}{10} & \cos x_3 \cdot \left(\cos\left(\frac{x_1}{10}\right) - \sin\left(\frac{x_1}{10}\right)\right) - \frac{1}{10}\sin x_3. \end{array} \quad (27)$$

From (13), we get that the desired angular velocity should be

$$\hat{\omega} \;=\; -\left(\text{sawtooth}(\theta - \text{atan2}(b, a)\right) - \frac{(b \cdot \dot{a} - a \cdot \dot{b})}{a^2 + b^2}. \quad (28)$$

Now, since the true angular velocity is $\dot{\theta} = -\rho_2 v \sin 2u_1$, we take

$$u_1 = -\frac{1}{2}\arcsin\left(\tanh\left(\frac{\hat{\omega}}{\rho_2 v}\right)\right). \quad (29)$$

The saturation function *tanh* is needed since the rudder cannot respond to any required $\hat{\omega}$. Indeed, if our controller ask to turn too fast for the boat, $\frac{\hat{\omega}}{\rho_2 v}$ will be more than 1, and the rudder can only do its best. The behavior of our controller is illustrated by Figure 4, where the sailboat has to slalom tightly between doors. We can see that the trajectory follows exactly the sine path (magenta).

The Python source codes associated to the simulation can be found at:

https://www.ensta-bretagne.fr/jaulin/slalompy.zip

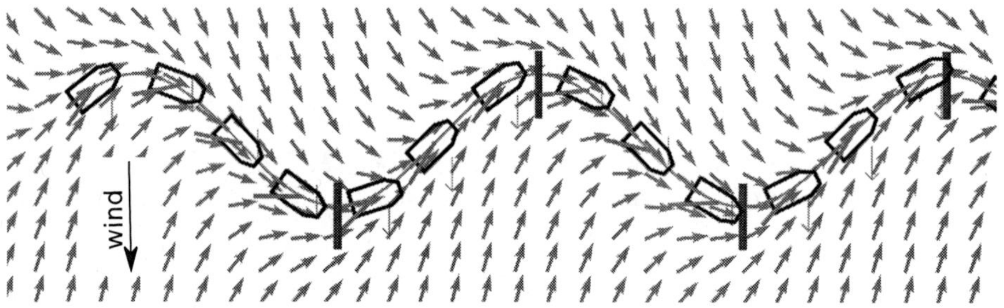

Figure 4: The sailboat robot slaloms through the blue doors

4 Conclusion

In this paper, we have proposed a new controller for sailboat robots which allows to take into account the curvature of the required field in order to anticipate as much as possible the required trajectory. To our knowledge, this is not considered by existing controllers (Le Bars and Jaulin, 2013) which are devoted to straight lines (Plumet et al., 2018). It has been shown that the required vector field could be followed exactly. This anticipation is crucial if we want to maneuver quickly and precisely as needed when we want to avoid an obstacle. This has been illustrated on a simulated test-case where a tight slalom is performed by a sailboat robot.

References

Cruz, N. and Alves, J. C. (2008). Ocean sampling and surveillance using autonomous sailboats. In *Proceedings of the 1st International Robotic Sailing Conference.*

Holger, K., Roland, S., Karim, J., and Mellinger, D. K. (2009). AAS endurance: An autonomous acoustic sailboat for marine mammal research. In *2nd International Robotic Sailing Conference.*

Isidori, A. (1995). *Nonlinear control systems: An Introduction, 3rd Ed.* Springer-Verlag.

Jaulin, L. (2015a). *Automation for robotics.* John Wiley & Sons.

Jaulin, L. (2015b). *Mobile robotics.* Elsevier.

Jaulin, L. and Le Bars, F. (2013). An interval approach for stability analysis: Application to sailboat robotics. *IEEE Transactions on Robotics*, 29(1):282–287.

Khatib, O. (1986). Real-time obstacle avoidance for manipulators and mobile robots. In *Autonomous robot vehicles*, pages 396–404. Springer.

Le Bars, F. and Jaulin, L. (2013). An experimental validation of a robust controller with the vaimos autonomous sailboat. In *Robotic Sailing 2012*, pages 73–84. Springer.

Miller, P. H., Hamlet, M., and Rossman, J. (2012). Continuous improvements to usna sailbots for inshore racing and offshore voyaging. In *5th International Robotic Sailing Conference*, pages 49–60. Springer.

Neumann, T. and Schlaefer, A. (2012). Feasibility of basic visual navigation for small sailboats. In *5th International Robotic Sailing Conference*, pages 13–22.

Pêtres, C., Romero-Ramirez, M.-A., and Plumet, F. (2011). Reactive path planning for autonomous sailboat. In *15th International Conference on Advanced Robotics (ICAR)*, pages 112–117. IEEE.

Plumet, F., Briere, Y., and Le Bars, F. (2018). Les voiliers robotisés. (ref. article : s7815). fre.

Schmitt, S., Le Bars, F., Jaulin, L., and Latzel, T. (2016). Obstacle avoidance for an autonomous marine robot—a vector field approach. In *Quantitative monitoring of the underwater environment*, pages 119–131. Springer.

Stelzer, R., Proll, T., and John, R. I. (2007). Fuzzy logic control system for autonomous sailboats. In *Fuzzy Systems Conference, 2007. FUZZ-IEEE 2007. IEEE International*, pages 1–6. IEEE.

Adaptive Probabilistic Tack Manoeuvre Decision for Sailing Vessels

Sébastien Lemaire
sebastien.lemaire@soton.ac.uk

Yu Cao
Yu.Cao@soton.ac.uk

Thomas Kluyver
thomas@kluyver.me.uk

Daniel Hausner
dh4n16@soton.ac.uk

Camil Vasilovici
crv1g16@soton.ac.uk

Zhong-yuen Lee
leezhongyuen@gmail.com

Umberto José Varbaro
ujv1u16@soton.ac.uk

Sophia M. Schillai
sms4g13@soton.ac.uk

University of Southampton
Boldrewood Innovation Campus
Southampton, S016 7QF

Abstract

To move upwind, sailing vessels have to cross the wind by tacking. During this manoeuvre distance made good may be lost and especially smaller vessels may struggle to complete a tack in averse wind and wave conditions. A decision for the best tack manoeuvre needs to be made based on weather and available tack implementations.

This paper develops an adaptive probabilistic tack manoeuvre decision method. The order of attempting different tacking strategies is based on previous success within a timeout, combined with an exploration component. This method is successfully demonstrated on the 1m long sailing vessel Black Python. Four strategies for crossing the wind were evaluated through adaptive probabilistic choices, and the best was identified without detailed sensory knowledge of the actual weather conditions.

Based on the positive results, further improvements for a better selection process are suggested and the potential of using the collected data to recognise the impact of weather conditions on tacking efforts is recognised.

In: S. M. Schillai, N. Townsend (eds.): Proceedings of the International Robotic Sailing Conference 2018, Southampton, United Kingdom, 31-08-2018

1 Introduction

Current work on manoeuvre planning for sailing robots focuses on long term piloting from planning the actions until the next waypoint to routing a vessel over longer distances (Tynan, 2018; Langbein et al., 2011). When the routing towards a waypoint, all sailing vessels rely on manoeuvres to cross the wind: tacking, where the vessel turns with the bow facing towards the wind, or jibing, where the vessel turns with the bow facing away from the wind. Typically, sailing robots are controlled by using a rudder to control the heading of the vessel and then adjusting the sails based on the relative wind direction (Gomes et al., 2017) even if recent sail designs may include heading control in the sail (Augenstein et al., 2017). Vessel speed and boat drag are often recognised as a factor for successfully completing a tack (Jouffroy, 2009a; Jouffroy, 2009b). (Cruz and Alves, 2014) recognises that a sailing robot can get stuck facing into the wind, in sailing terms 'in irons', and suggests to recover from such a situation by increasing speed; thus gaining rudder control through letting the sails loose so the wind can push the boat backwards. (Tranzatto et al., 2015) studies how to perform fast and smooth tack manoeuvres using control system theory and compares 3 different rudder controls for tacking, however tack fails are not mentioned. Modeling a tacking manoeuvre could also be done to analyse the problem of tack failure. Several studies developed tacking simulators based partially on experimental measurements (Masuyama and Fukasawa, 2011; Roncin and Kobus, 2004; Spenkuch et al., 2010), however they are applied on large sailing boats where tack failure is not an issue, and is hence not discussed. When sailing the 1m long Southampton Sailing Robot, the Black Python, we found that the success and speed of a tack manoeuvre for such a small vessel not only depends on having sufficient speed and suitable rudder action to pass through the wind, but also on passing through the wave fronts pushed towards it by the wind. Where a human sailor would make choices about the sail and rudder settings based on speed and wave observations and experience, making small adjustment as the manoeuvre proceeds, implementing this process for a robotic sailor is challenging. Not only is it difficult to translate experience into software, but also the amount of sensor data that is required increases significantly compared to a dead-reckoning tack manoeuvre. As an alternative to fully measuring and considering all factors involved, the introduced system adapts to the wind and waves conditions by testing and evaluating several available tack methods.

This paper investigates a dynamic weighting approach to choose the best method to perform a tack in order to minimise the number of failed tacking attempts whilst having minimal sensor knowledge of weather and boat state. After introducing the Black Python vessel, its sensors and the software structure in the Systems section, we focus on the software components that control the tack manoeuvre, suggesting several tacking implementations and a tack weighting process based on previous successful and failed tack manoeuvres. The methods introduced are demonstrated on experiment results obtained in coastal waters near Southampton.

2 System presentation

2.1 The boat

The Black Python, see Figure 1a, is a one-meter-long Lintel mono-hull sailing robot yacht of class IOM (International One Metre) designed by David Creed. This boat is designed for racing performances and is used in remote controlled regatta. Three different sets of sails with a sail area of 6000, 4100 and 2700 cm^2 can be used depending on the wind conditions. The hull's beam is 165 mm and the hull displacement is 4000 g. Minor changes have been made to fit wiring. The Black Python uses bulb keel that is 420 mm deep, and has a spade rudder for steering. Profile view of the Black Python is shown in Figure 1b.

2.2 Electronics

A Raspberry Pi 3 B (RPi) microcomputer is used as the main control board. It is powered by a 5V USB power bank. A foam stand and plastic box keep the RPi away from water that occasionally gets inside the boat.

A uBLOX MAX M8Q GPS unit gives the boat position and velocity, it communicates with the RPi via I^2C (Inter-Integrated Circuit). A conventional USB WiFi dongle is placed together with the GPS and a small IMU (Pololu AltIMU-10 v4) unit on the top of the mast. Inside the boat, an Xsens MTi 3 IMU (Inertial Measurement Unit) is used. It includes an accelerometer, gyroscope and magnetometer. The IMU is placed directly on top of the RPi. The yacht uses a custom made wind vane (Figure 1c). Two magnets are attached to the rotating part and a Pololu AltIMU-10 v4 is placed on the stator. The magnetometer on the IMU detects the change in the magnetic field as the wind vane is rotated by the wind, determining the wind direction relative to the boat.

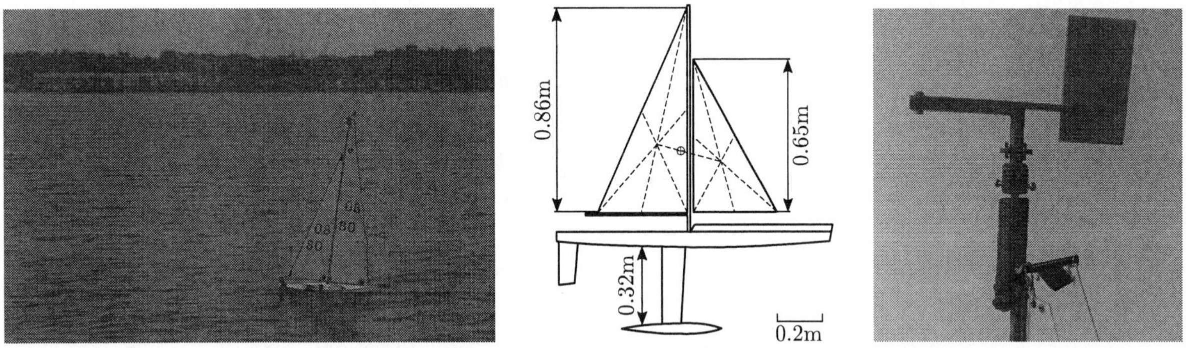

(a) The Black Python in Southampton water

(b) Profile view showing the center of effort of the smallest set of sails (Papadopoulos, 2018)

(c) Custom made wind vane

Figure 1: The Black Python

A Futaba S3003 servomotor is used to drive the rudder and a HiTec 785 HB winch servomotor drives both mainsail and jib at once. A remote control receiver as well as a multiplexer are used to take control of the sailing robot in case of emergency or during the launch and recovery phases. The motors, multiplexer and RC receiver are powered by 4 AA batteries which are monitored with an Adafruit INA219 current sensor to ensure sufficient power for a remote controlled recovery.

2.3 Software

As mentioned in the previous section, the main computer of the Black Python is a Raspberry Pi. It runs the GNU/Linux distribution Ubuntu 16.04. The software is written in Python and utilises ROS (Robot Operating System[1], (Quigley et al., 2009)); a framework for writing robot software. ROS includes a collection of tools and libraries to simplify the task of developing complex behaviours. In the ROS ecosystem, the code is structured around scripts called nodes. Each node can send messages under a certain topic name: this is called *publishing*. Nodes can also listen for specific topics by *subscribing* to them. The entire software developed by the Southampton Sailing Robot Team is made available under the MIT free software licence[2].

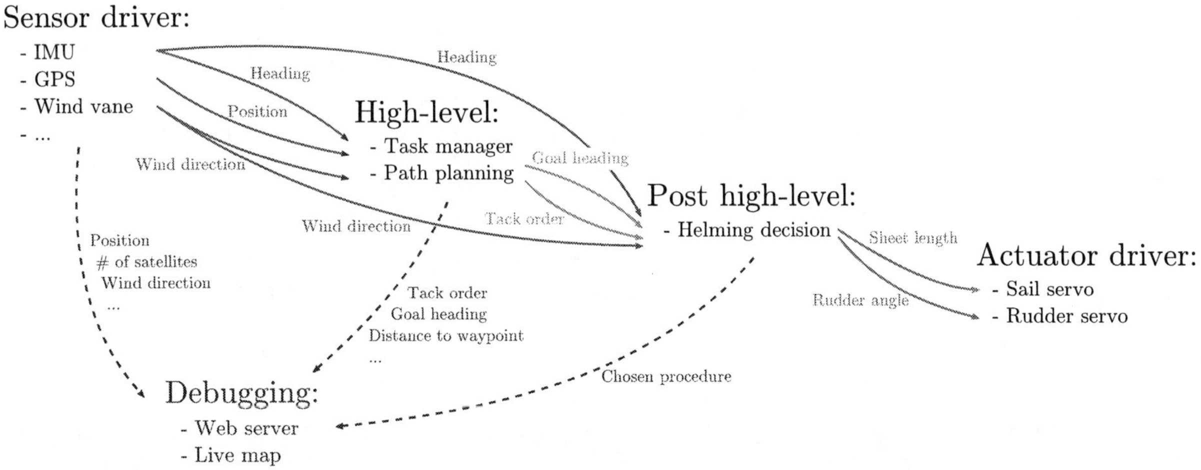

Figure 2: Structure of the code, arrows symbolise ROS messages passing

[1]`https://ros.org`

[2]`https://github.com/Maritime-Robotics-Student-Society/sailing-robot`

The software of the Black Python is structured as follows:

- *Drivers*: nodes talking directly to hardware components; this includes reading sensor data from the GPS, wind vane or obstacle avoidance camera as well as adjusting servo motor positions to set rudder angle and sail sheet.
- *High-level*: path planning nodes deciding when to switch tack and which heading to follow to complete a task. Available tasks are: reaching a waypoint, keeping a position, and avoiding an obstacle.
- *Post high-level*: the *helming node* converting the goal heading or tack order from high-level nodes into rudder angle and sail position. The *helming node* is described in depth in section 3.
- *Debugging*: nodes for visualisation of messages; from maps with waypoint and boat positions to graphical displays of angle information like heading, goal heading, and wind direction

Figure 2 illustrates the software structure, giving key nodes and the messages exchanging information between them. To achieve a particular assignment, the user configures the robot via parameter files. These include a list defining the tasks to run and task specific parameters like waypoint coordinates.

3 Post high-level: dynamic tack control

Weather conditions are sometimes not suitable for tacking on such small boats (dues to waves for example). In this situations a reliable approach is to jibe instead. The switch between tacking and jibing to cross the wind was implemented in the past on the Black Python using a user defined parameter set before each test. Whilst being a fail-safe method, we measured that jibing instead of tacking makes the vessel looses about 3m made good, when beating at 50 degree from the wind and with a speed of 0.75m/s. This leads to a loss of about 6 seconds per jibe, hence jibing should only happen when necessary. The post high-level was introduced to increase the choice of available manoeuvres alternatives to jibing and to automate the decision between all available tack and jibe manoeuvres.

The post high-level layer currently consists of the *helming node* alone, which operates between the low level drivers and the high level nodes. When the boat is not trying to switch tack, the rudder angle is set using a heading PID control. To set the sail sheet length, a predefined look up table containing the apparent wind direction versus the sail sheet length is used. When a tack order is given from the high-level nodes, the *helming node* is responsible for implementing a successful tack by changing the sheet length and rudder angle demands, choosing from a set of available procedures based on past events and a dynamic weight system with an exploration component.

In the context of the *helming node*, a procedure is the name given to a series of instructions that commands the sail and the rudder to perform a change of tack (jibe or tack). Several procedures are implemented:

- Basic tack (`BasicTack`): the rudder is set to its maximum position either on the port side if the boat was sailing on a port tack, or on the starboard side otherwise.
- Basic jibe (`BasicJibe`): the rudder is set to its maximum position in reversed compared to a tack. To help with bearing away the sails are sheeted out.
- Tack with sheet out (`TackSheetOut`): the rudder is set to its maximum position to perform a tack, and the sails are slightly sheeted out. On a conventional sailing boat the main sail tends to make the boat go more upwind, when the jib pushes the boat to go downwind. Sheeting out the jib can help with tacking, however on the Black Python both sails share the same control. This procedure hence tries to reduce the power in the jib by sheeting out a little both sails while conducting the tack.
- Tack with speed build up (`TackIncreaseAngleToWind`): to speed up the boat and gain momentum to aid passing the tipping point of the tack the boat will bear away for 5 seconds at 80 degrees from the wind. It will then perform an usual tack with setting the rudder to its maximum position.

The basic functioning of the *helming node* is as follows: Before each switch of tack, the procedure list is ordered by the time taken by each procedure in the past. The procedures in the list are tried in order until one succeeds to make the boat switch tack before a user defined timeout. After each procedure attempt, the time it took is recorded for future use to determine the order in the procedure list. A tack procedure is considered a success if the time the procedure took in order to have the boat on the opposite tack (at an angle between 50 and 120 degrees relative to the wind) is bellow the user defined `timeout`. If a procedure fails it is placed further towards the end of the list by recording a value of 1.5 times the timeout. To ensure that all procedures are attempted, an exploration coefficient is considered as well.

Three user defined variables are used:

- `timeout`: timeout in seconds after which a procedure is considered as failed
- `ProcedureList`: initial order of the procedure list
- `Exploration coefficient`: probability (0 to 1) of picking an untried procedure instead of the top list entry

The `ProcedureList` is a python list of dictionaries, each element of the list has three dictionary keys: `Procedure` which is a pointer to the procedure class, `TimeList` a list of the time taken by the last 10 attempts of this procedure, and finally `InitPos` the initial position of the procedure in the `ProcedureList` as defined by the user.

Every time a tack is attempted, the `ProcedureList` is ordered based on weight. The procedure with the lowest weight will be tried first. The weights are computed as follows:

If the procedure have been tried in the past (ie. `TimeList` is not empty), its weight is the mean of the elements of the `TimeList`, in other words the average time the procedure took in the past. On the other hand, if the procedure has never been tried before, the `TimeList` is empty, then the `Exploration coefficient` is used. To know if the procedure will be picked by the exploration, a random number between 0 and 1 is generated. If it is above the `Exploration coefficient` no exploration is done. A combination of the `timeout` and the `InitPos` is given as the weight. This places the procedure between already attempted procedures, before the failed ones but after the succeeded ones whilst keeping all unused procedures in order of the initial list. Otherwise, if the randomly picked value is bellow the `Exploration coefficient` the procedure is placed at the top of the list by giving it a random weight between 0s and 0.1s (the random value ensures an arbitrary selection between methods picked by the exploration coefficient). The computation of the weight is summarised in Algorithm 1.

Algorithm 1 Function to get the weight of each procedure

```
function GETWEIGHT(procedure)
    if procedure.TimeList not empty then
        return mean(procedure.TimeList)
    else
        if random(0,1) < explore_coef/number_of_untested_procedures then
            return random(0, 0.1)
        else
            return timeout + 0.01*procedure.InitPos
```

Once the `ProcedureList` is ordered, it will not be reordered until the next high-level command to change tack. The list entries are attempted in order until a procedure successfully completes before the timeout. If all elements of the `ProcedureList` are tried without a success, the procedure selection will continue with the same list, beginning at the top entry.

3.1 Example of run

In this section a step by step example of a fictional run is described and illustrated in figure 3. The wind is coming from the North, and the boat is beating upwind. The user defined parameters are as follows:

- `Timeout`: 15s
- `ProcedureList`: [`BasicTack`, `TackSheetOut`, `BasicJibe`]
- `Exploration coefficient`: 0.3

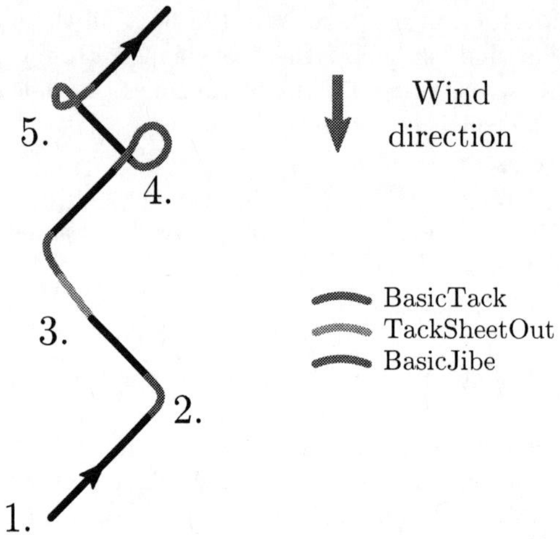

Figure 3: Fictional example of run with the *helming node* procedure picking

1. The boat starts sailing on a port tack, close hauled.

2. The *helming node* receives a tack change command from the high-level nodes. The `ProcedureList` is sorted. None of the unused (all) procedures is moved to the top of the list through the `Exploration coefficient`, hence the `ProcedureList` is as defined by the user: [`BasicTack`, `TackSheetOut`, `BasicJibe`]. The `BasicTack` is tried, it succeeds in 7 seconds. The boat continues on a starboard tack.

3. The high-level commands the *helming node* to change tack. The `ProcedureList` is ordered. This time the weight of the `BasicTack` is 7, making it the node with the fastest average time. The `Exploration coefficient` places the `TackSheetOut` procedure first in the `ProcedureList`: [`TackSheetOut`, `BasicTack`, `BasicJibe`]. The `TackSheetOut` is tried, but does not succeed before the timeout of 15 s. A value of 22.5 (1.5 times the timeout) is stored in the `TimeList` for the `TackSheetOut`. The next procedure on the list, the `BasicTack`, is tried and succeeds in 8 seconds. The boat continues on a port tack.

4. The high-level commands the *helming node* to change tack. The `ProcedureList` is ordered, the `Exploration coefficient` causes no re-ordering. The `ProcedureList` is: [`BasicTack`, `BasicJibe`, `TackSheetOut`]. The `BasicTack` is tried and fails, the next procedure (`BasicJibe`) is tried and succeed in 9 seconds. The boat continues its course.

5. The high-level commands the *helming node* to change tack. The `ProcedureList` is ordered, no re-ordering is caused by the `Exploration coefficient`. The `BasicTack` has a weight of 12.5 (mean of 7, 8 and 22.5), `TackSheetOut` has a weight of 22.5 (failed once) and `BasicJibe` has a weight of 9. The order is hence [`BasicJibe`, `BasicTack`, `TackSheetOut`]. The `BasicJibe` is tried and it is a success.

3.2 Discussion on parameter selection

The user defined parameters were purposefully kept at a minimum and chosen to have an easily understandable meaning. The exploration coefficient may however demand some experience of the boat behaviour to be set properly. A low exploration coefficient should be used when the user is confident with his ordering of the procedure list or when he knows that very few tacks will be performed during the test. Hence finding the best possible manoeuvre is not as rewarding as finding a working manoeuvre. On the other hand when a larger number of tacks are to be expected, setting the exploration coefficient higher will help finding the most performant method. The timeout should be set at the minimum value that allows ones boat to perform a tack manoeuvre in the expected (or all) weather conditions. A too low timeout will lead to the helming node cycling through the procedures and keeping failing when a too high value will make the boat loose time when trying a procedure for the first time.

4 Experiments

The experiment was conducted on the sea near Southampton Sailing Club shown in Figure 4. Wind direction gradually shifted from south-east to south. The averaged wind direction according to the wind direction data collected from the wind vane on the boat is shown in Figure 4a. Averaged wind speed was 4 knots with gusts at 5 knots. Small waves with height of 15 - 20 cm have been observed during the test.

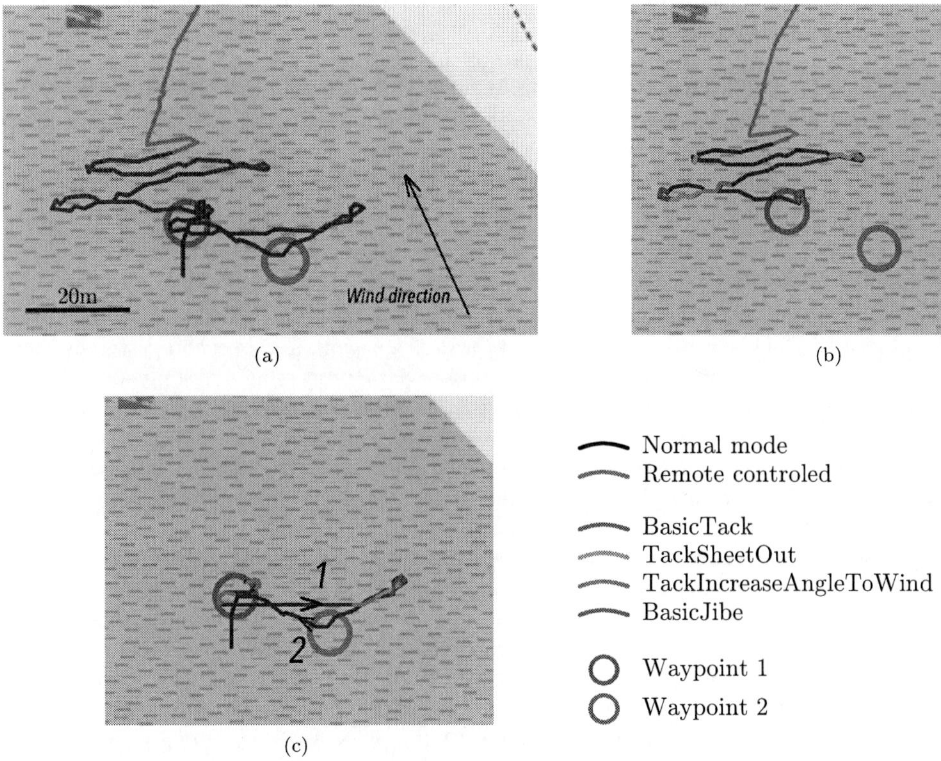

Figure 4: Experiment results in Southampton water

During the experiment, the boat was programmed to sail between two waypoints separated by 20m shown as a green and red circle on Figure 4. The Black Python was released from a runway on north of the visible map. To avoid being washed back by the waves the boat was first navigated into the ocean by a remote control. Once the boat was further away from the shore, the autonomous mode was activated. It tried to reach the first waypoint using the *helming node* described in previous section. An acceptance radius of 1.5m was set for all waypoints, parameters used in this experiment were:

- `timeout`: 30s
- `ProcedureList`: [`BasicTack`, `TackSheetOut`, `TackIncreaseAngleToWind`, `BasicJibe`]
- `Exploration coefficient`: 0.3

Experiment results are shown in Figure 4. In the first leg, three tack manoeuvres were made to reach the waypoint. As shown in Figure 4b, the first tack was done with a `TackSheetOut` procedure. Whilst being the second element in the initial `ProcedureList`, it was tried first because during the manual controlled phase the *helming node* was still running and a `BasicTack` failed, moving this procedure at the end of the list. This behaviour is not intended and will be fixed. The `TackSheetOut` succeeded in 9s. For the second manoeuvre, `TackSheetOut` was tried first and failed, `TackIncreaseAngleToWind` was then conducted and succeeded in 19s. For the next tack, `TackIncreaseAngleToWind` was tried first. It however failed, the next element in the sorted list now being `TackSheetOut`. This procedure was tried and also failed, finally a `BasicJibe` was conducted with success in 19s.

After getting to the first waypoint, the boat sailed towards the second waypoint as shown in Figure 4c. Two `TackIncreaseAngleToWind` were performed and succeeded. For the next tack manoeuvre, all four procedures were tried: first `TackIncreaseAngleToWind` did not manage to perform the tack leading to the boat bearing away, then `TackSheetOut` was tried without success. Later `BasicJibe` was tried, the boat managed to switch tack, however, the jibe did not finish on time and switching to `BasicTack` was needed to finalise the manoeuvre. The last leg from the second waypoint to the first one is downwind, the *helming node* did not start any procedure and the boat sailed in a straight line. After the boat reached the first waypoint again, wind condition stopped us from doing any further repeating tests.

On this day, the low wind speed made it particularly difficult leading to a lot of failed manoeuvres. Here the exploration never picked a random untried procedure because all procedures were tried in the first three tacks. With such low wind conditions, increasing the timeout might help reducing the number of failed manoeuvres because regardless of the method the boat is very slow to switch heading. This test demonstrated good functioning of the *helming node* and results show that in such weather jibing or taking up speed by bearing away first helps switching tack.

5 Concluding remarks and future work

Tacking on a small sailing boat is a delicate manoeuvre, the method presented here efficiently identifies a good strategy to perform a tack. The key aspect of the *helming node* framework and decision system is its simplicity; it does not rely on any additional hardware or even complex data processing but only on sensors that are already widely used on robotic sailing boats (heading and wind direction). It makes it easy to implement and to debug. From an user point of view, the simplicity of the implementation is also visible by the limited number of parameters and their physical meaning. Only 3 user defined parameters are needed (a timeout, a sorted list of procedure and an exploration coefficient) and all of them have an easy to understand meaning, no extensive knowledge of the boat behaviour or the weather conditions is needed.

The *helming node* system was successfully demonstrated, but it can still be improved on several aspects. If all the manoeuvres of the `ProcedureList` fail, this can mean that the selected timeout is too short. Instead of rerunning the list with the same parameter, increasing the timeout automatically would be judicious. Additional tack procedures could further consider the sea state, for example by timing the tacks based on the position of the boat on the waves. Although some work to detect wave period has been done by the Southampton Sailing Robot Team, more tests are still needed to refine the method and integrate it as a procedure. Lastly, for now only the time taken by a procedure is measured to assess its performances. To more accurately consider the distance lost during the jibe manoeuvre combining the time with the distance gained towards the next waypoint, for example, could be an improvement of the *helming node* weighting process. Also the method is currently not suitable for long term tests where weather conditions might change over time. An improvement of the weighting that includes an aging parameter would be preferable in this case.

After performing more tests with this new system, a better understanding of each procedure will be gained and more precise and general conclusions concerning the best way to perform a tack in a specific weather condition can be drawn.

References

Augenstein, T., Singh, A., Miller, J., Pomerenk, A., Dean, A., and Ruina, A. (2017). Using a controlled sail and tail to steer an autonomous sailboat. In *Robotic Sailing 2016*, pages 91–103. Springer.

Cruz, N. A. and Alves, J. C. (2014). Navigation performance of an autonomous sailing robot. In *Oceans-St. John's, 2014*, pages 1–7. IEEE.

Gomes, L., Costa, A., Fernandes, D., Marques, H., and Anjos, F. (2017). Improving instrumentation support and control strategies for autonomous sailboats in a regatta contest. In *Robotic Sailing 2016*, pages 45–56. Springer.

Jouffroy, J. (2009a). A control strategy for steering an autonomous surface sailing vehicle in a tacking maneuver. In *Systems, Man and Cybernetics, 2009. SMC 2009. IEEE International Conference on*, pages 2391–2396. IEEE.

Jouffroy, J. (2009b). On steering a sailing ship in a wearing maneuver. *IFAC Proceedings Volumes*, 42(18):26–31.

Langbein, J., Stelzer, R., and Frühwirth, T. (2011). A rule-based approach to long-term routing for autonomous sailboats. In *Robotic Sailing*, pages 195–204. Springer.

Masuyama, Y. and Fukasawa, T. (2011). Tacking simulation of sailing yachts with new model of aerodynamic force variation during tacking maneuver. *SNAME Journal of Sailboat Technology*, 1.

Papadopoulos, I. (2018). Techniques to improve the tacking performance of model scale robot yachts. Technical report, University of Southampton.

Quigley, M., Conley, K., Gerkey, B., Faust, J., Foote, T., Leibs, J., Wheeler, R., and Ng, A. Y. (2009). ROS: an open-source robot operating system. *ICRA workshop on open source software*, 3(3.2):5.

Roncin, K. and Kobus, J.-M. (2004). Dynamic simulation of two sailing boats in match racing. *Sports Engineering*, 7(3):139–152.

Spenkuch, T., Turnock, S., Scarponi, M., and Shenoi, A. (2010). Real time simulation of tacking yachts: how best to counter the advantage of an upwind yacht. *Procedia Engineering*, 2(2):3305–3310.

Tranzatto, M., Liniger, A., Grammatico, S., and Landi, A. (2015). The debut of aeolus, the autonomous model sailboat of ETH zurich. In *OCEANS 2015-Genova*, pages 1–6. IEEE.

Tynan, D. (2018). An attractor/repellor approach to autonomous sailboat navigation. In *Robotic Sailing 2017*, pages 69–79. Springer.

Isobath Following using an Altimeter as a Unique Exteroceptive Sensor

Luc Jaulin
ENSTA Bretagne
LabSticc
lucjaulin@gmail.com

Abstract

We consider an underwater robot equipped with an altimeter (a simple echo-sounder oriented downward), a barometer and a low cost gyroscope. The robot has no compass. We show that using vector field control combined with a Kalman observer, it is possible to follow an isobath (which is a curve that connects all points having the same depth). The robustness of the controller is validated on a simulation.

1 Introduction

This paper deals with the difficult problem for a robot to explore an unknown environment, without any localization system and without being lost. By being lost, we mean not being able to reach a target set, or equivalently not being able to come back home. Under the water, we can easily know the depth using a barometer and the problem of finding a path can be considered in the horizontal 2D plane. If we are able to measure some quantities such as the altitude, the temperature, the salinity, etc, we can find a reliable path which allows us not being lost. This the case of underwater animals such as marine turtles, or whales which follow isotherms (Lohmann and Lohmann, 1996) with the help of an internal compass. These underwater animals do not know where they are but they know from the evolutionary process that if they follow a sequence of iso-potentials they will perform a cycle which is stable.

Here, to perform the exploration we will rely on a bathymetric approach. This is motivated by the thesis of Rohou (Rohou, 2017) who has shown that once some underwater exploration has been done, a bathymetric localization can be performed if we know the map. Better than that, an efficient and reliable bathymetric simultaneous localization and mapping (SLAM) could be done using the tube approach of Rohou (Rohou, 2017). Now, Rohou assumed that the mission was already performed and an external localization system had to be used for this purpose. This motivates the need to explore an unknown underwater environment using an altimeter as the single exteroceptive sensor. The paper proposes a solution for this exploration following an isobath in a simple manner, with simple low cost sensors, without any compass and without surfacing to use the GPS.

In: S. M. Schillai, N. Townsend (eds.): Proceedings of the International Robotic Sailing Conference 2018, Southampton, United Kingdom, 31-08-2018

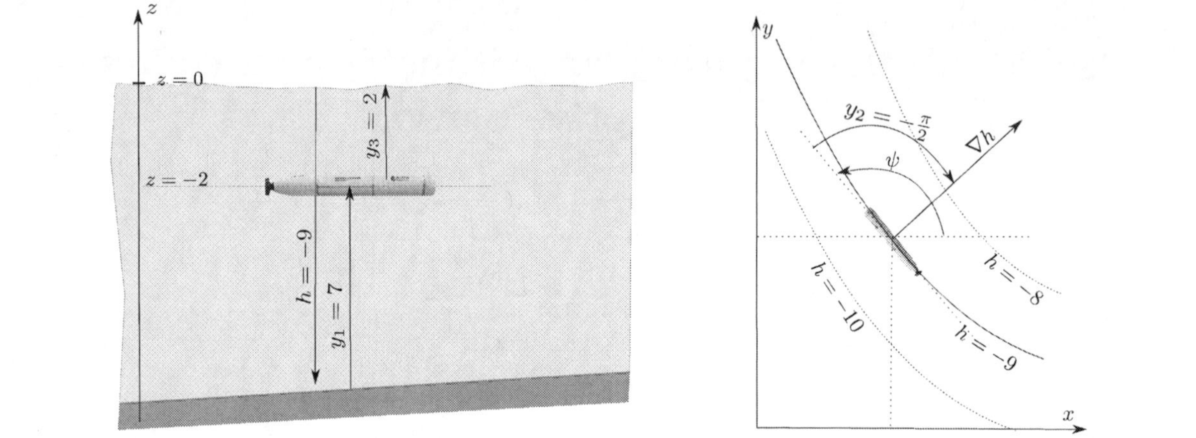

Figure 1: Underwater robot that has to follow an isobath

2 Problem

An isobath is an imaginary curve that connects all points having the same depth $h(x, y)$ below the surface, *i.e.*, an underwater level curve. Consider an underwater robot (Jaulin, 2015a) described by the state equation:

$$\left\{\begin{array}{lll} \dot{x} & = & \cos\psi \\ \dot{y} & = & \sin\psi \\ \dot{z} & = & u_1 \\ \dot{\psi} & = & u_2 \end{array}\right. \tag{1}$$

where (x, y, z) corresponds to the position of the robot and ψ is its heading angle. The robot is able to measure its altitude y_1 with an echo sounder which is a simple and low-cost sonar transmitting a sound pulse. The time interval between emission and return of a pulse is recorded and provides the distance to the seafloor. In the first part of the paper, we assume that we measure the angle y_2 of the gradient of h in its own frame. This assumption which is not always realistic will be lifted later in Section 4. Moreover the robot is able to know its depth y_3 using a pressure sensor.

The observation function of our system is thus

$$\left\{\begin{array}{lll} y_1 & = & z - h(x, y) \\ y_2 & = & angle(\nabla h(x, y)) - \psi \\ y_3 & = & -z \end{array}\right. \tag{2}$$

where $\nabla h(x, y)$ the gradient of h. For instance, if $\mathbf{y} = \left(7, -\frac{\pi}{2}, 2\right)$, the robot knows that it is following an isobath corresponding to $-y_1 - y_3 = -7 - 2 = -9$m, at a depth of 2m. This is illustrated by Figure 1.

3 Controller

In this section, we propose a controller of the form $\mathbf{u} = \mathbf{r}(\mathbf{y})$, which makes the robot follows an isobath corresponding to $h_0 = -9$m at a depth $\overline{y}_3 = 2$m. For the control of the depth, we can take a proportional control of the form

$$u_1 = y_3 - \overline{y}_3. \tag{3}$$

For the heading, assume that we first want to follow the isobath just below the robot. Equivalently we want to the robot be perpendicular to $\nabla h(x, y)$, for instance, $y_2 = \pm\frac{\pi}{2}$, depending if we want the gradient on our right or on our left. Take for instance $e_1 = y_2 + \frac{\pi}{2}$ as an error. This means that we want to have an angle $\overline{y}_2 = -\frac{\pi}{2}$ with ∇h, or equivalently, we want ∇h on the right, as in Figure 1. If $e_1 = 0$, we follow an isobath, but is may not be the right one. We thus have to consider another error corresponding to $e_2 = -y_3 - y_1 - h_0 = 0$. It $e_2 = 0$, we are just above the right isobath but maybe not parallel to it. If both $e_1 = 0$ and $e_2 = 0$, we are on the right isobath $(e_1 = 0)$ but we also go on the right direction $(e_2 = 0)$. For the heading, we may take the following controller

$$u_2 = \tanh(e_2) + \text{sawtooth}(e_1) = -\tanh(h_0 + y_3 + y_1) + \text{sawtooth}(y_2 + \frac{\pi}{2}), \tag{4}$$

where *tanh* creates a saturation (Jaulin, 2015b). The *sawtooth* function is given by:

$$\text{sawtooth}(\widetilde{\theta}) = 2\text{atan}\left(\tan\frac{\widetilde{\theta}}{2}\right) = \text{mod}(\widetilde{\theta} + \pi, 2\pi) - \pi \tag{5}$$

As illustrated by Figure 2, the function corresponds to an error in heading. The interest in taking an error $\widetilde{\theta}$ filtered by the *sawtooth* function is to avoid the problem of the $2k\pi$ modulus: we would like a $2k\pi$ to be considered non-zero.

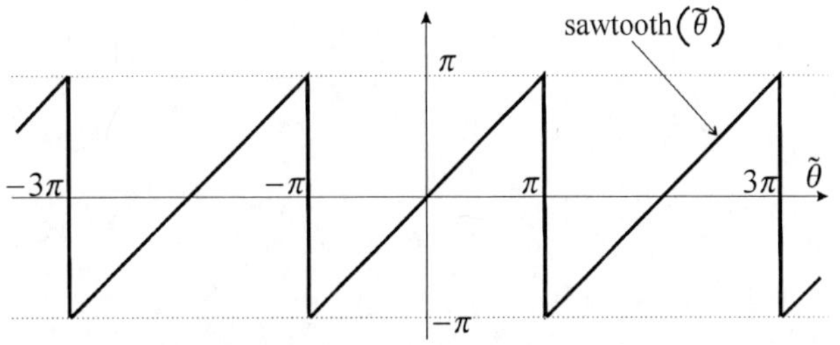

Figure 2: *Sawtooth* function used to avoid the jumps in the heading control

The controller may thus be given by

$$\mathbf{u} = \begin{pmatrix} y_3 - \overline{y}_3 \\ -\tanh(h_0 + y_3 + y_1) + \text{sawtooth}(y_2 + \frac{\pi}{2}) \end{pmatrix}. \tag{6}$$

The coefficients for the controller (all taken here equal to ± 1) should be tuned in order to have the stability and correct time constants.

Remark. For the heading, the controller is close to a proportional and derivative control, where $\tanh(h_0 + y_3 + y_1)$ corresponds to the proportional term and $\text{sawtooth}(y_2 + \frac{\pi}{2})$ to the derivative term. For the heading control, we could take a proportional and derivative control of the form

$$u_2 = (\overline{y}_1 - y_1) + \dot{y}_1 = (\overline{y}_1 - y_1) + \dot{z} - \nabla h(x, y) \cdot \begin{pmatrix} \cos\psi \\ \sin\psi \end{pmatrix}, \tag{7}$$

where $\overline{y}_1 = -\overline{y}_3 - h_0$, $\dot{z}$ can be assumed to be zero and $\nabla h(x, y) \cdot \begin{pmatrix} \cos\psi \\ \sin\psi \end{pmatrix}$ is assimilated to $\text{sawtooth}(y_2 + \frac{\pi}{2})$. Recall that x, y, ψ are not measured and cannot be used by our controller.

Test-case

We consider a seafloor described by

$$h(x, y) = 2 \cdot e^{-\frac{(x+2)^2 + (y+2)^2}{10}} + 2 \cdot e^{-\frac{(x-2)^2 + (y-2)^2}{10}} - 10. \tag{8}$$

For the initial condition, we take $x = 2, y = -1, z = -2, \psi = 0$ we obtain the trajectory depicted on Figure 3.

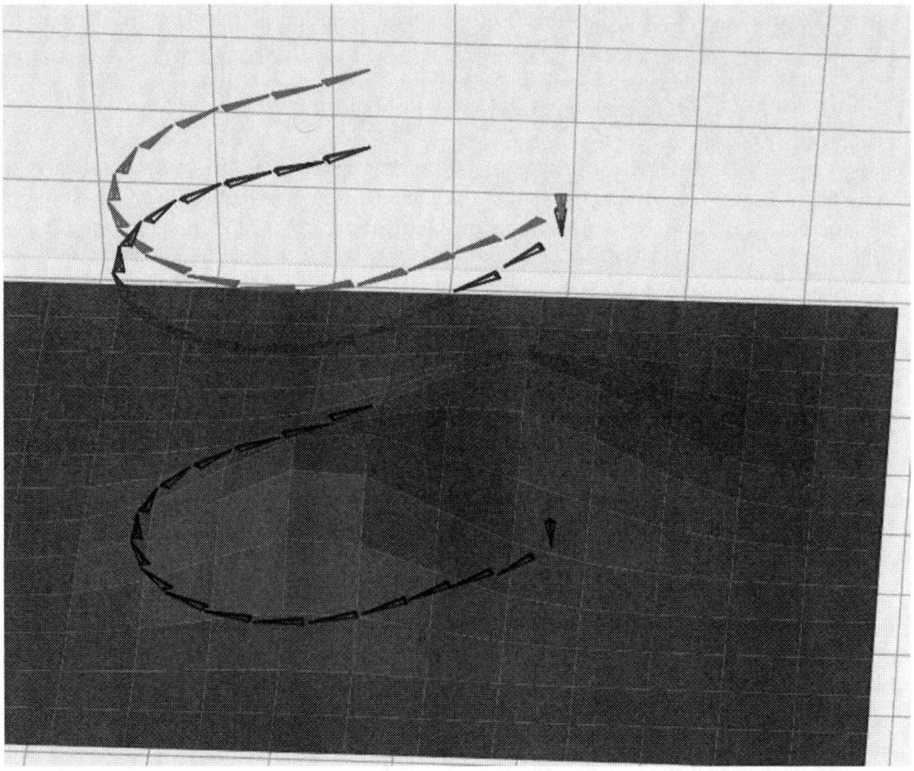

Figure 3: Simulation of underwater robot (blue) following an isobath. The surface shadow (gray) and the seafloor shadow are also painted.

4 Using an observer to get the gradient of the seafloor

We assumed previously that the robot was able to measure the vertical distance y_1 to the seafloor using an echo-sounder. Now, we have also assumed that the robot was also able to measure the angle y_2 of the underneath isobath which is not realistic with a low-cost sonar. The angle $y_2 = angle(\nabla h(x,y)) - \psi$ should thus be estimated using an observer such as a Kalman filter. The idea is similar to what is proposed in (Jaulin, 2015a) in the context where a car has to follow a wall at a given distance from measuring the distance to the wall only.

In the local frame of the robot projected onto the surface, the underneath plane satisfies the equation

$$z_1 = p_1 x_1 + p_2 y_1 + p_3 \tag{9}$$

where (p_1, p_2) corresponds to the gradient and p_3 to $h(x,y)$. Let us note that

- (i) Since the robot has no compass, it has no idea of its orientation and can thus have an estimation of its neighborhood only in its own frame. This is why we have chosen to express the plane in the robot frame.
- (ii) Without limitation of the method, we have chosen a linear model. Now, for more accuracy, we could have taken a model of the seafloor with more parameters, such as a quadratic model $z_1 = p_1 x_1^2 + p_2 y_1^2 + p_3 x_1 y_1 + p_4 x_1 + p_5 y_1 + p_5$. This quadratic model is particularly interesting if the seafloor is smoothly curved.

Prediction. We assumed that the underneath seafloor is locally planar. We thus have:

$$\dot{\mathbf{p}} = \begin{pmatrix} 0 & \dot{\psi} & 0 \\ -\dot{\psi} & 0 & 0 \\ 1 & 0 & 0 \end{pmatrix} \mathbf{p}. \tag{10}$$

This equation can be understood by the fact that the gradient (p_1, p_2) turns with the robot and that the variable p_3 increases when the robot moves with the gradient (i.e., with p_1). As a consequence, we can assume the following prediction equation for the underneath plane:

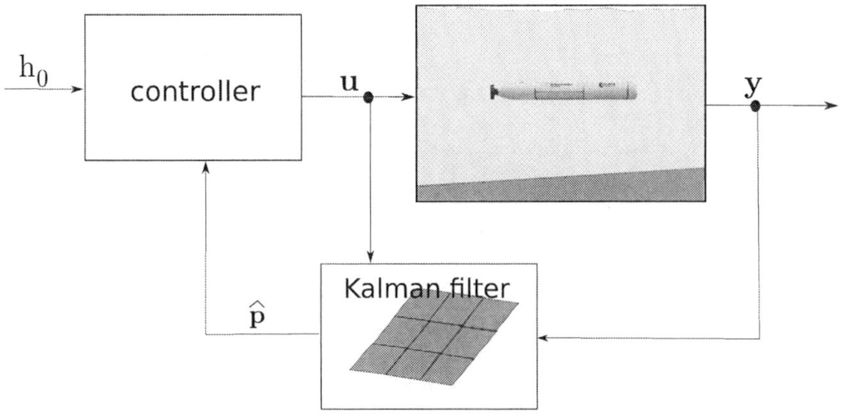

Figure 4: Use of a Kalman filter to estimate the seafloor

$$\mathbf{p}(k+1) = \begin{pmatrix} 1 & dt \cdot u_2(k) & 0 \\ -dt \cdot u_2(k) & 1 & 0 \\ dt & 0 & 1 \end{pmatrix} \mathbf{p}(k) + \boldsymbol{\alpha}(k). \tag{11}$$

where $\boldsymbol{\alpha}(k)$ is a white Gaussian noise, the covariance matrix of which depends on how planar is the seafloor.

Correction. Since we measure both the altitude y_1 and the depth y_3, we have an indirect measurement of p_3. We thus have the correction equation:

$$-y_1 - y_3 = \begin{pmatrix} 0 & 0 & 1 \end{pmatrix} \mathbf{p}(k) + \beta(k). \tag{12}$$

where $\beta(k)$ is a white Gaussian noise. The unknown gradient can thus be estimated by a Kalman filter which returns an estimation $\hat{\mathbf{p}}$ of $\mathbf{p}$. This is illustrated by Figure 4. In the Kalman filter box, we have represented a plane which what is actually estimated by the Kalman observer.

The controller (6) may thus be given by

$$\mathbf{u} = \begin{pmatrix} y_3 - \overline{y}_3 \\ -\tanh(h_0 - \hat{p}_3) + \text{sawtooth}(\text{atan2}(\hat{p}_2, \hat{p}_1) + \frac{\pi}{2}) \end{pmatrix}. \tag{13}$$

As illustrated by Figure 5, the trajectory oscillates and the isobath following is less accurate. If we observe the covariance matrix for $\mathbf{p}$, we observe that when the seafloor becomes planar for a while, the estimation is good at the beginning and becomes slowly very bad. This is due to the fact that a straight trajectory corresponds to a singularity. Now, in such a case, the robot goes ahead and performs the isobath following. Later, when the seafloor changes its orientation, the estimation of $\mathbf{p}$ becomes more accurate and the robot is able to change its orientation accordingly.

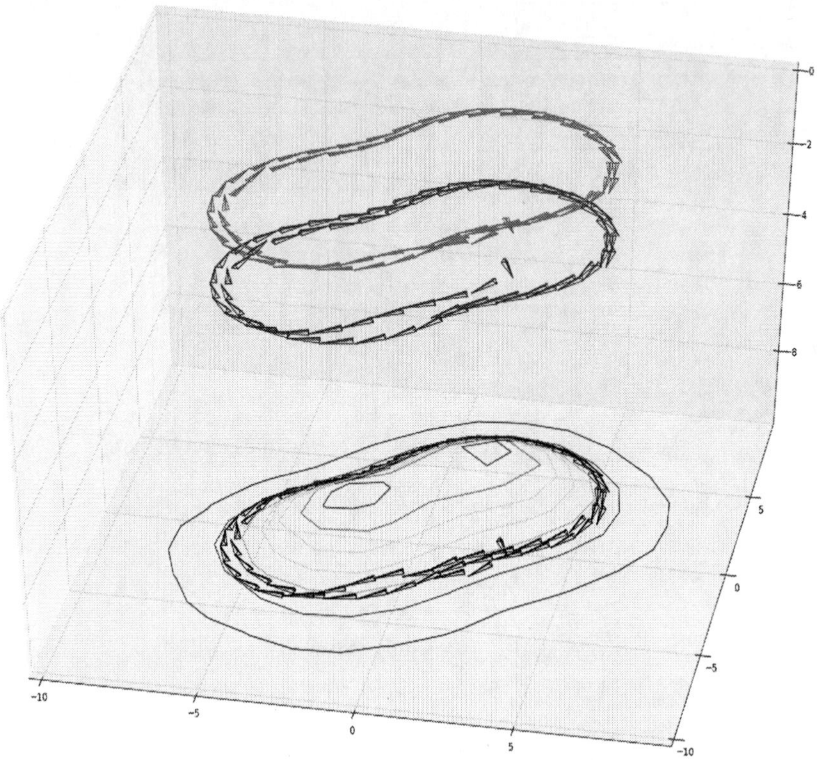

Figure 5: The underwater robot follows an isobath, without compass and with a low-cost echo-sounder

5 Conclusion

In this paper, we have shown that it was possible to follow an isobath with low-cost sensors that are not greedy in energy. Now, once we are able to follow such an isobath, it should be possible to detect the existence of a cycle using proprioceptive sensors (motors for instance) (Aubry et al., 2013). Such a cycle could thus lead us to a localization and also to the possibility to perform a bathymetric SLAM. Since the control is bathymetric, we have all elements to solve a pure bathymetric *explore and return* problem (Newman et al., 2002) which has not been solved yet, to our knowledge.

References

Aubry, C., Desmare, R., and Jaulin, L. (2013). Loop detection of mobile robots using interval analysis. *Automatica*, 49(2):463–470.

Jaulin, L. (2015a). *Automation for robotics.* John Wiley & Sons.

Jaulin, L. (2015b). *Mobile robotics.* Elsevier.

Lohmann, K. and Lohmann, C. (1996). Orientation and open-sea navigation in sea turtles. *Journal of Experimental Biology*, 199(1):73–81.

Newman, P., Leonard, J., Tardós, J. D., and Neira, J. (2002). Explore and return: Experimental validation of real-time concurrent mapping and localization. In *Robotics and Automation, 2002. Proceedings. ICRA'02. IEEE International Conference on*, volume 2, pages 1802–1809. IEEE.

Rohou, S. (2017). *Reliable robot localization: a constraint programming approach over dynamical systems.* PhD thesis, Brest.